What Does the Moon Smell Like?

151 ASTOUNDING SCIENCE QUIZZES

EVA EVERYTHING

ECW PRESS

Published by ECW Press
2120 Queen Street East, Suite 200,
Toronto, Ontario, Canada M4E 1E2
416.694.3348 / info@ecwpress.com

LIBRARY AND ARCHIVES CANADA CATALOGUING IN PUBLICATION

Everything, Eva
What does the moon smell like? : 151 astounding science quizzes / Eva Everything.

ISBN 978-1-55022-822-9

1. Science — Miscellanea. i. Title.

Q173.E918 2008 500 C2007-907098-1

Editor: Crissy Boylan
Cover and Text Design: Tania Craan
Author Photo: Biserka
Typesetting: Mary Bowness
Production: Rachel Brooks
Printing: Thomson-Shore

This book is set in Helvetica and Fairfield
and printed on paper that is 30% post consumer recycled.

The publication of *What Does the Moon Smell Like?* has been generously supported by the Government of Ontario through the Ontario Book Publishing Tax Credit; by the OMDC Book Fund, an initiative of the Ontario Media Development Corporation; and by the Government of Canada through the Book Publishing Industry Development Program (BPIDP).

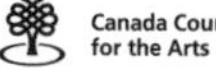

Canadä

PRINTED AND BOUND IN THE UNITED STATES

TO THOSE WHOM THIS BOOK INSPIRES

&

IN MEMORY OF MY MOTHER

Acknowledgements

Thanks to all the "mad" scientists (and I mean that in the best sense of the word "mad") both present and past, whose amazing discoveries and ideas are like chocolate for my brain; to the brilliant scientists who generously contributed their time and knowledge; to the smartest and funniest people I know for their advice and insights regarding this book; to my friends (who are also among the smartest and funniest people I know) for putting up with me all these years, for encouraging me and always aiding and abetting my endeavors; to my husband and parents for their love and support; to the Ontario Arts Council Writers' Reserve; and to the publisher, editor, and the entire ECW Press team who made it possible for you to read these words.

Table of Contents

Introduction

When I asked the smartest, funniest people I know whether they'd rather read a popular science book of prose, or a quiz book, most of them said quiz book. They thought it would be more fun and interesting than a regular book. Well, fun is my middle name, and I couldn't resist the challenge of coming up with a quiz book that had more to offer your brain than other quiz books.

Your brain is amazing and it deserves more than the usual 500 standard questions, and 500 one-line answers. It deserves to be challenged, tickled, and entertained by the mind-boggling details of science, everyday life, technology, and nature.

What Does the Moon Smell Like? aims to induce an enjoyable brain state with 151 deluxe, challenging quizzes, suitable for all ages, presented in a new and improved format.

Along with the question, four choices, and an answer, you get two bonus paragraphs that make things more interesting. The first one is before the question, and it gives you the context in which the question is being asked. The second paragraph expands on the answer. Within a few minutes, you've learned a handful of surprising, weird, and/or wonderful, wacky, amazing, intriguing, or unusual facts. No pain, all gain.

While your brain zips along, you can give your thumbs a rest. Instead of going to the back of the book for the answers, after reading a question and deciding which choice you think is correct, just turn the page and then flip the book to read the answer. When you're ready for the next question, simply flip the book back. It's so easy, even an adult can do it.

But before you start turning the pages, and the book, one last thing: have you noticed how quickly science and technology are advancing? Breakthroughs happen on a daily basis, and the body of knowledge just keeps gaining weight and growing. The information in this book was accurate when I wrote it. If some things have changed by the time you read this, don't blame me. Blame the scorching pace of discovery and innovation!

I hope you enjoy reading *What Does the Moon Smell Like?* as much as I enjoyed writing it.

Eva Everything
www.thebraincafe.ca

Q

On the morning of April 18, 1955, Albert Einstein's brain was removed. He had just died, so he didn't mind. His brain was being preserved for future study, which was a great idea — in theory. But, in reality, Einstein's brain couldn't be studied, because it disappeared, and no one, not even his kin, knew where it was. The world's most famous brain was missing for decades. Then, in 1978, Einstein's lost brain was found.

Who tracked down Einstein's missing brain?

a) a detective

b) a pathologist

c) a reporter

d) Einstein's granddaughter

Einstein's Brain — Lost

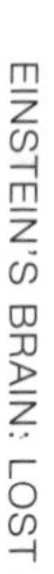

Who tracked down Einstein's missing brain?

a) a detective

b) a pathologist

c) a reporter

d) Einstein's granddaughter

CORRECT ANSWER:

c) a reporter

When you think about who is most likely to solve a medical mystery, pathologists, detectives, or even a dedicated granddaughter, might come to mind. But it was Steven Levy, then a reporter for a magazine called *New Jersey Monthly*, who tracked down Einstein's brain. Where on earth had Einstein's brain been all those years? Well, let's see . . .

Einstein's Brain — Found

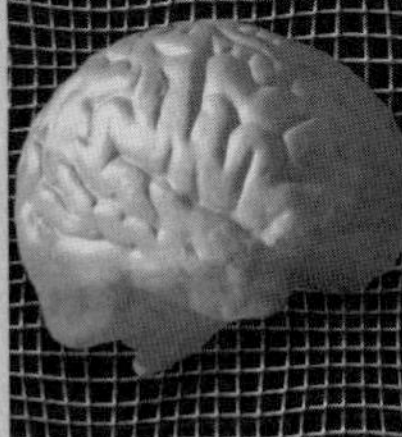

Q

Einstein's brain was missing for decades, and you have to wonder why it took so long to find, because locating the genius lobes turned out to be a no-brainer. Einstein once said something like, "Everything should be made as simple as possible, but not simpler," and it's exactly what the successful sleuth did.

Where was Einstein's brain found?

a) in a freezer in a lab at Oxford University, U.K.

b) in a margarine tub at a brain bank in Hamilton, Canada

c) in a pair of pickling jars in a home office in Wichita, U.S.

d) in a private medical collection in Tokyo, Japan

Einstein's Brain — Found

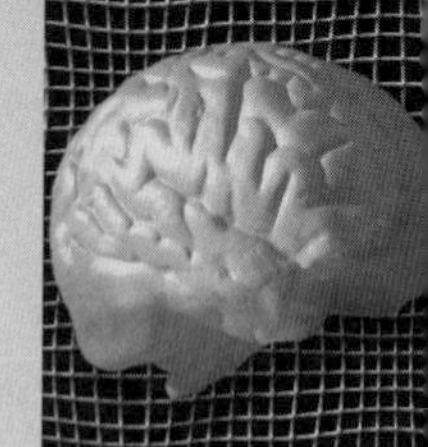

Where was Einstein's brain found?

a) in a freezer in a lab at Oxford University, U.K.

b) in a margarine tub at a brain bank in Hamilton, Canada

c) in a pair of pickling jars in a home office in Wichita, U.S.

d) in a private medical collection in Tokyo, Japan

CORRECT ANSWER:

c) in a pair of pickling jars in a home office in Wichita, U.S.

Steven Levy took the simplest approach and tracked down the pathologist who'd removed Einstein's brain back in 1955. Dr. Thomas Harvey had since retired and moved to Wichita, Kansas. At first, the doctor denied knowing anything about Einstein's brain, but the young reporter wouldn't take no for an answer. Finally, Dr. Harvey broke down and admitted that he had the brain, right there, in the very office in which they were sitting. He went over to a box labelled "Costa Cider" and pulled out two large pickling jars containing Einstein's lobes. Seeing the brain that had changed the world was huge for Steven Levy. He said that "it was almost a religious experience."

Einstein's Groovy Brain

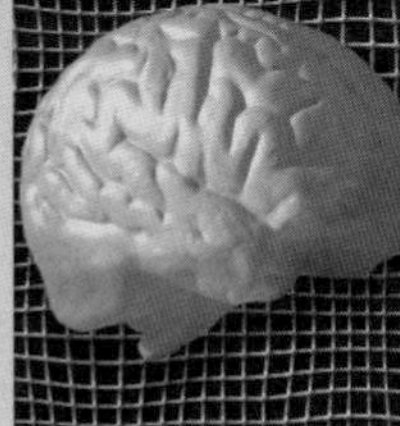

Q

When the news broke that Einstein's brain had been found, scientists were excited. Have you ever wondered if Einstein's brain was special? That's exactly what the world's top brain researchers wanted to know too. Einstein's brain was well preserved, but well past the thinking stage, so it couldn't be tested in action, or take an IQ test. But his brain's anatomy was just begging to be explored.

How does Einstein's brain compare to the average brain? It's . . .

a) bigger
b) heavier
c) longer
d) wider

Einstein's Groovy Brain

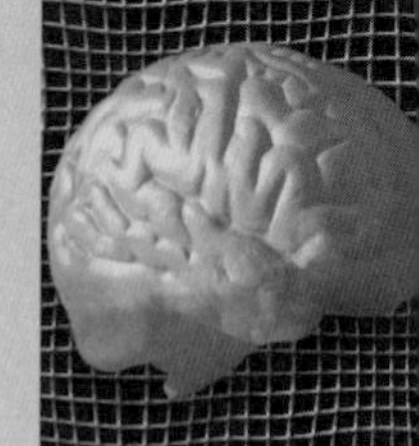

How does Einstein's brain compare to the average brain? It's . . .

a) bigger
b) heavier
c) longer
d) wider

CORRECT ANSWER:

d) wider

If you have an average male brain, it's as long as Einstein's, but yours might be heavier. Before you get too excited, I should tell you that brain size relates to body size. An average-sized male's brain weighs about 1,400 grams (3 lb). Einstein was smaller than average and his brain, at 1,230 grams (2.7 lb), weighs less too. It's not bigger, longer, or heavier than average, but it is 15% wider, and has unusual grooves in the areas for math and three-dimensional thinking. Did these differences contribute to Albert Einstein's genius? No one really knows. But we do know that there's more to intelligence than brain size or anatomy. That's why, at this very moment, researchers all over the world are trying to unravel the mysteries of the human mind in all its glory. Pull up a chair, it could take a while.

Q

Duct tape devotees will tell you that duct tape is like The Force — it has a dark side and a light side, and it binds the universe together. Has anyone told the astrophysicists? Think of how much time they'd save solving the riddles of the universe if they only knew! Here on Earth, we've found a universe of uses for duct tape in the more than 65 years that it's been around. Did you know that duct tape was originally called duck tape? Do you know why?

Why was duct tape called duck tape? Because . . .

a) it was invented by Dr. Donald "Duck" Mallard

b) it was mispronounced by the military and the name stuck

c) it was waterproof like a duck

d) the manufacturer's logo featured a duck

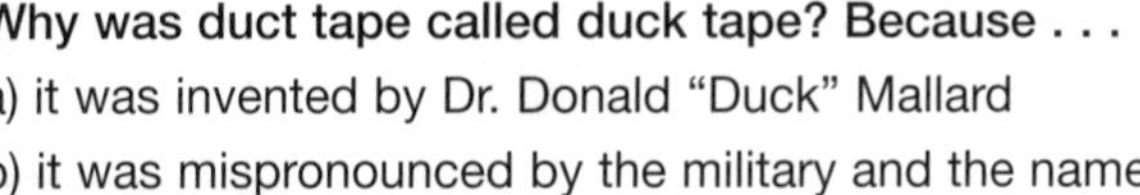

The Binding Force of the Universe

MEET ME @ The BrainCafé

Why was duct tape called duck tape? Because . . .

a) it was invented by Dr. Donald "Duck" Mallard
b) it was mispronounced by the military and the name stuck
c) it was waterproof like a duck
d) the manufacturer's logo featured a duck

CORRECT ANSWER:

c) it was waterproof like a duck

The U.S. military needed tape that would keep moisture out of its ammunition cases. In 1942, during World War II, a company that made medical tape supplied an army green, cloth-backed, rubber-adhesive, waterproof tape. Since water rolled off it (like water rolls off a duck's back), it was nicknamed duck tape. The military found all kinds of uses for it. After the war, builders used the tape to connect the ducts of home heating and air conditioning systems. The name morphed from duck to duct tape, and the classic silver colour was introduced to match the colour of the ducts.

To Squish and to Unsquish

Have you ever wondered why yellow sticky notes stick, and then unstick? The secret is in the glue. It's made up of tiny, round balls called microcapsules. When you press sticky notes down, you flatten the tiny glue balls, so there's more glue in contact with the surface to which you're attaching your note. When you pull the note off, the microcapsules unsquish, and with less glue in contact with the surface, it comes up easily. Sticky notes were a great invention, but why would anyone come up with the idea for a note that sticks and unsticks?

What was the creator of the original yellow sticky note doing when he came up with the idea?

a) cataloguing books in a library
b) organizing his tax return
c) reconstructing a crime scene
d) singing in a choir

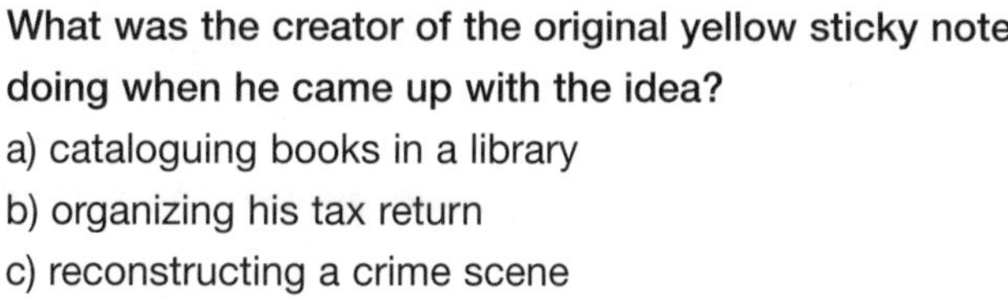

A

What was the creator of the original yellow sticky note doing when he came up with the idea?

a) cataloguing books in a library
b) organizing his tax return
c) reconstructing a crime scene
d) singing in a choir

CORRECT ANSWER:

d) singing in a choir

Art Fry enjoyed singing in the choir, but was frustrated by his bookmark. It kept falling out, and he kept losing his page. Then, one day, while singing, he had a revelation — the solution to his problem was a repositionable note, one that stuck to the page, but could be easily removed. Luckily, he just happened to be a product development engineer, and had access to the perfect glue for the job. It had been invented by another scientist years before, but no one had found a use for it. Fry put the adhesive on the back of a scrap of paper, and his bookmark dilemma was solved, not to mention that he'd just created the now commonplace sticky note.

Slippery Pans

MEET ME @ The Brain Café

Once upon a time, our ancestors had to use gobs of grease to keep their eggs from sticking to the frying pan. But even so, food often stuck, and then they had to apply lots of elbow grease to clean up the burned-on mess. Non-stick pans, coated with Teflon®, changed all that. Foods didn't stick, even when fried without grease, and cleanup was a breeze. It was a great idea, and a real innovation in cookware, but like many inventions, non-stick pans started out as a happy accident in the lab. The inventor wasn't trying to make a slippery pan. He was trying to make something else altogether.

What was the inventor of Teflon® trying to make?

a) a lubricant

b) a refrigerant

c) an adhesive

d) an edible oil product

Slippery Pans

What was the inventor of Teflon® trying to make?

a) a lubricant
b) a refrigerant
c) an adhesive
d) an edible oil product

CORRECT ANSWER:
b) a refrigerant

Chemist Ray Plunkett was researching new refrigerants on his very first project for a chemical company in 1938. In one experiment, he expected to find a refrigerant gas, but got a weird white powder instead. Luckily, he decided to study its properties, and it turned out be amazing stuff — heat resistant, very slippery, and non-reactive with most chemicals. First used by the Manhattan Project to make atom bombs during World War II, it was later harnessed for peaceful purposes, like making non-stick pans.

Q

Although the search is on for alternative fuels, most of the cars in today's world are still powered by gasoline. When we need more fuel, we drive to the gas station and fill up. Once upon a time, in the earliest days of motoring, there were no gas stations. Where did pioneering motorists get gas?

Where did motorists go for gas in the earliest days?

a) car dealership

b) general store

c) oil refinery

d) pharmacy

Got Gas?

Where did motorists go for gas in the earliest days?

a) car dealership
b) general store
c) oil refinery
d) pharmacy

CORRECT ANSWER:
d) pharmacy

Getting fuelled up was a challenge back then. More than 110 years ago, gasoline, or petrol, was used mostly as a cleaning agent, and it was only stocked in small quantities by pharmacies. Road trips had to be planned carefully to avoid running out of gas. The earliest motorists were adventurous types, and they were up for the challenge, even if they were considered to be nutty, eccentric, or worse. For the most part, motoring was considered to be a hobby for the rich. Back then, few people took cars seriously, or realized their potential.

The World's First Car Race

More than 130 years ago, when the race was on to build the first true automobile, everyone (and his brother) was building some kind of car. Horses, who powered transportation at the time, were terrified of the noisy, motorized monsters, and so were many people. Some even thought that cars were the work of the devil, but resistance was futile. The development of the automobile was inevitable. In 1894, a Paris newspaper, *Le Petit Journal*, organized a reliability test to see which cars performed the best on a 130-kilometre (80 mile) drive from Paris to Rouen, France. Twenty cars set out to prove their worth.

What was the top speed of the winner?

a) 8 km/h (5 mph)

b) 24.5 km/h (15 mph)

c) 54 km/h (33.5 mph)

d) 80.5 km/h (50 mph)

What was the top speed of the winner?

a) 8 km/h (5 mph)

b) 24.5 km/h (15 mph)

c) 54 km/h (33.5 mph)

d) 80.5 km/h (50 mph)

CORRECT ANSWER:

b) 24.5 km/h (15 mph)

As soon as the drivers hit the road, the reliability test turned into a race, though not a car race as we know it. Each driver had a mechanic with him and, apparently, the racers stopped for lunch along the way. Most of the cars that passed the finish line had Daimler engines, including the two cars judged to be the winners, one made by the Peugeot brothers, and the other a Panhard. Newspapers all over the world covered the race, and Daimler got the kind of publicity that money can't buy. Racing had a huge impact on the popularity of cars. People liked to watch, and it wasn't long before they wanted to drive.

A Different Kind of Engine

Q

Rudolf Diesel set out to build a different kind of engine — one that didn't need a spark, or electricity, to run. He spent more than a decade developing his revolutionary engine, and finally unveiled it at the 1900 World's Fair in Paris. Aside from the engineers and motoring buffs who came to see his engine, members of the general public were also drawn to Diesel's demonstrations: they followed their noses.

What smell drew the public to Diesel's demonstrations?

a) Diesel fuel

b) french fries

c) gunpowder

d) perfume

A Different Kind of Engine

What smell drew the public to Diesel's demonstrations?

a) Diesel fuel
b) french fries
c) gunpowder
d) perfume

CORRECT ANSWER:

b) french fries

Diesel was using peanut oil in his engine, and the combusting oil smelled enough like french fries to attract people looking for food. Some of them must have been disappointed, but the hungry engineers and motor fans in the crowd were amazed. No one had ever seen an engine running on straight peanut oil before. Petroleum-based fuels were expensive at the time, and Dr. Diesel envisioned his engines running on cheaper, easier-to-get, vegetable oils. More than 100 years after Diesel's demonstration, about half the vehicles in Europe are diesel, but instead of vegetable oils, they run on a fuel made from petroleum. Diesel invented the engine that bears his name, but not the fuel, which was concocted long after he'd died.

Q

When the afternoon munchies hit, you might find yourself craving a fresh, hot, crunchy snack. While the microwave pops the kernels, imagine a Stone Age family, squatting around a cooking fire long ago — at least 10,000 years ago. Do you feel the heat of the fire? Do you smell the woodsy smoke? Do you hear a familiar sound, so eerily similar to the popping coming from your microwave? They may not have had microwaves, but Stone Age people did have popcorn, and they knew how to pop it. The Aboriginal peoples of North America were the world's first popcorn lovers.

What did Aboriginal peoples do with popcorn?

a) make popcorn beer

b) make popcorn soup

c) perform a ceremonial dance with popcorn garlands on their heads

d) all of the above

The World's First Popcorn Lovers

What did Aboriginal peoples do with popcorn?

a) make popcorn beer
b) make popcorn soup
c) perform a ceremonial dance with popcorn garlands on their heads
d) all of the above

CORRECT ANSWER:

d) all of the above

Popcorn probably originated in what is now Mexico, and spread from there. Many native peoples ate it as a snack. Some made popcorn soup, and some even brewed popcorn beer! Popcorn garlands adorned the heads of Aztec maidens during ceremonial dances, and decorated the statues of their gods. Some people were so into it, they were even buried with it. One-thousand-year-old popcorn kernels, so well preserved that they still could have been popped, have been found in tombs in Peru. Now there's a snack to die for.

Your Candy Bar Eating Style

Q

If you like candy bars, you probably find yourself reaching for one in the afternoon or early evening. That's the most popular time of day to indulge in a chocolate bar. Why? Common sense says that it could be because we need a little extra fuel to get us through to dinner. Even though many candy bar lovers eat them at around the same time of day, *how* we go about indulging depends on our eating style. Seriously. According to a Canadian survey . . .

Women are twice as likely as men to eat a chocolate bar by . . .

a) breaking it into smaller pieces before eating it
b) taking progressively bigger bites, saving the biggest bite for last
c) using a knife and fork
d) wolfing it down in a few big bites

Your Candy Bar Eating Style

A

Women are twice as likely as men to eat a chocolate bar by . . .

a) breaking it into smaller pieces before eating it
b) taking progressively bigger bites, saving the biggest bite for last
c) using a knife and fork
d) wolfing it down in a few big bites

CORRECT ANSWER:

a) breaking it into smaller pieces before eating it

Apparently, some women have a love/guilt relationship with chocolate, so they break off a small piece to eat and plan to save the rest for later. As if! According to Krystyna Sieciechowicz, an anthropologist who studies our relationship with food, women think that breaking up the bar is a more feminine way to indulge, and some do it because they like to savour each little morsel, slowly, one after the other. Not so for the men in the survey: two out of five reported wolfing their bar down in just a few big bites. Many males also like to take progressively bigger bites, saving the biggest bite for last. Eating a candy bar with a knife and fork is a very rare style found mostly among England's upper class, or upper class wannabes.

An Equal Opportunity Snack

Q

More women than men are crazy about chocolate, but potato chips are an equal opportunity snack, craved by females and males alike. In English-speaking countries, they're one of the most popular snacks by any other name. Translation? French fries are called chips, and chips are called crisps in England. It's even more confusing in Australia, New Zealand, and South Africa, where both chips and fries are simply called chips. You're probably thinking, "Yeah, yeah, I know that." But what else do you know about potato chips, or crisps?

Who invented potato chips?

a) a mother who loved her deep fryer
b) a scientist working for a food giant
c) an ornery chef making fun of a customer
d) an unemployed actor living at home with his parents

An Equal Opportunity Snack

Who invented potato chips?

a) a mother who loved her deep fryer

b) a scientist working for a food giant

c) an ornery chef making fun of a customer

d) an unemployed actor living at home with his parents

CORRECT ANSWER:

c) an ornery chef making fun of a customer

The story goes that potato chips were invented about 150 years ago at a posh resort, Moon Lake Lodge, in Saratoga Springs, New York, by a chef bent on payback. When a fussy diner dared to send back his fried potatoes, claiming that they were too thick and soggy, the ornery chef decided to serve up just what the customer ordered. He fried paper-thin spud slices in oil, salted them, and voila — potatoes so crispy they shattered at the touch of a fork! To the chef's surprise, the ridiculously thin, crisp, potato "chips" were a huge hit with the fussy diner and his friends. Back then, potato chips were only served in restaurants. Bagged chips came about 75 years later.

Q

All the world loves a comedian . . . (drum roll) . . . except for the woman married to him! Alright, I made that up — the world doesn't necessarily love a comedian. Luckily, you don't have to be a comic to joke around. Who hasn't made an absolutely hilarious wisecrack at some time or other? There's nothing quite as satisfying as cracking wise, is there?

When someone makes a wisecrack, who laughs the most?

a) optimists
b) pessimists
c) teenagers
d) the wisecracker

Cracking Wise

When someone makes a wisecrack, who laughs the most?

a) optimists

b) pessimists

c) teenagers

d) the wisecracker

CORRECT ANSWER:

d) the wisecracker

If that's not the answer you expected, you're in good company — neither did psychologist Robert Provine. When he and his students observed 1,200 people going about their everyday lives, they were surprised to find that the wisecrackers laughed almost 50% more than their audiences. If you doubt it, the next time the cracks are flying, check out who's making the cracks, who's laughing the most, and for how long. You may be as surprised as Dr. Provine was.

LOL

Q

When you're alone, do you talk to yourself? It's okay to admit it. Many people do, and some even smile to themselves. But unless they're watching or listening to something funny, people rarely laugh out loud when no one else is there. The young adults in one study laughed 30 times more in the company of others than when they were alone. While socializing, the women laughed 126% more than the men. Why were they laughing so much? More importantly . . .

When did the young women laugh most? When they were . . .

a) embarrassed

b) with males they did not like

c) with males they liked

d) with their girlfriends

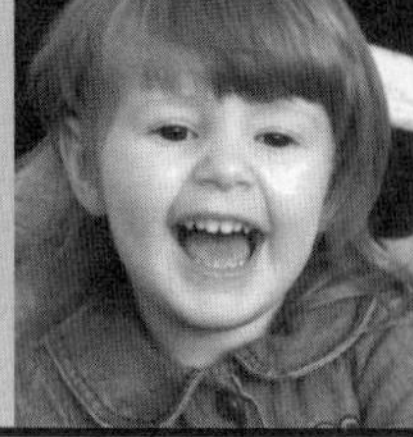

When did the young women laugh most? When they were . . .

a) embarrassed
b) with males they did not like
c) with males they liked
d) with their girlfriends

CORRECT ANSWER:

c) with males they liked

Women laugh more than men in general, but they laugh the most around men, especially men they like. The more a woman likes a man, the more she laughs. It's not because women are happier, or men are funnier. For some unknown reason, a man's role is to get laughs, and the woman's role is to laugh. This pattern pops up in culture after culture all over the world. Why? There's lots of speculation, but no one's sure. One thing we do know is that it starts early in life. Remember the class clown? Chances are the Bozo was a boy.

Cracking Up

Have you ever joined a joke in progress? You arrived too late to hear the punchline, but just in time to see your friends cracking up. Even though you didn't know what they were laughing at, did you start laughing too? You probably did. Laughter is contagious. So contagious, in fact, that there have even been epidemics of laughter. One laughing epidemic spread like wildfire in Africa in 1962, and claimed at least 1,000 victims.

How did the epidemic of laughter start? It began when . . .

a) a candid camera-type show aired a hilarious clip
b) a popular comedian coined a clever wisecrack
c) someone slipped something into a town's water supply
d) three girls at a boarding school got the giggles

Cracking Up

How did the epidemic of laughter start? It began when . . .

a) a candid camera-type show aired a hilarious clip
b) a popular comedian coined a clever wisecrack
c) someone slipped something into a town's water supply
d) three girls at a boarding school got the giggles

CORRECT ANSWER:

d) three girls at a boarding school got the giggles

The epidemic started with three giggling schoolgirls in what is now Tanzania. Soon, 95 students were howling. Two and a half months later, the school had to be closed because the students couldn't get a grip. But it didn't end there. When they got home, the students infected their families and friends. The epidemic spread from person to person, and from one community to the next, for at least half a year. Did the person who laughed last, laugh best? We'll never know.

Q

Lead pencils don't actually have lead in them. Pencil leads are made from a mixture of clay and a very soft, black, non-toxic mineral called graphite. The great, great, great, great grandaddy of modern pencils was developed after a deposit of very pure graphite was discovered more than 450 years ago.

What led to the discovery of the graphite deposit?

A . . .

a) dog digging for a bone

b) glacier receding

c) lake drying up

d) violent storm

Unleaded Lead

What led to the discovery of the graphite deposit?

A . . .

a) dog digging for a bone

b) glacier receding

c) lake drying up

d) violent storm

CORRECT ANSWER:

d) violent storm

After a violent wind storm uprooted oak trees near Borrowdale, England, local shepherds investigated the damage, and found mounds of black stuff where the trees had stood. They thought it was coal and tried to burn it, but it wouldn't fire up. They did find a use for it though. It was messy to handle, but good for marking their sheep. Before long, inventive types had cleaned up the mess by cutting the graphite into square pieces and encasing them in wood. Rumour has it that William Shakespeare used one of the earliest pencils made from Borrowdale graphite to write some of his plays and sonnets.

The Pencil Tree

Anyone who's ever sharpened a pencil knows that it smells of wood. What kind of wood? That depends on where your pencil was made. American-made pencils are usually cedar, Russian ones are white pine, and pencils made in China are often basswood. Have you ever wondered how many pencils you can get from a tree? I'll give you a hint — more than enough to give one to every person in your school, or workplace.

How many pencils does an average cedar tree yield? About the right number to give one to every person in . . .

a) Brasilia, Brazil

b) Canberra, Australia

c) Reykjavik, Iceland

d) Washington, DC, U.S.

The Pencil Tree

How many pencils does an average cedar tree yield? About the right number to give one to every person in . . .

a) Brasilia, Brazil
b) Canberra, Australia
c) Reykjavik, Iceland
d) Washington, DC, U.S.

CORRECT ANSWER:
c) Reykjavik, Iceland

An average cedar yields approximately 172,000 pencils, about enough to give one to every person in Reykjavik. In Iceland, as in the rest of the world, the yellow pencil rules. Most of the billions of pencils made in the world every year are painted yellow, but there are other colours available too. In Argentina and Brazil, you'll find black-painted pencils. In Australia and New Zealand, red ones are common; in Germany, there are green as well as blue ones. In the Nordic countries, unpainted pencils are a popular nod to environmentalism, though they haven't threatened the world dominance of the yellow pencil yet. The colours on the outside of the pencil may vary, but when it comes to the hardness of the lead, the world is united. The most common pencil lead on the planet is HB, which stands for hard and black.

Printing Like a Genius

Q

The first time you printed your name, you were probably using the writing implement favoured by some of the most brilliant minds of all time. Albert Einstein worked out his theory of relativity in pencil, and Thomas Edison carried one in his vest pocket for making notes. Author John Steinbeck went through 60 pencils a day while writing *The Grapes of Wrath*. Every time the point got dull, he picked up a fresh pencil. But what if you didn't have that option? Say you had reams of lined paper, a passion for writing, but just one average-sized, new pencil.

What's the maximum you could you write with an average pencil?

a) 10 average-length magazine articles

b) a 12-volume encyclopedia

c) the United Nations' Universal Declaration of Human Rights

d) this book

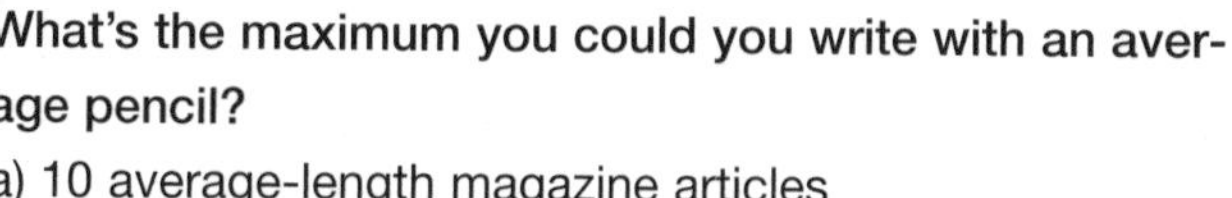

Printing Like a Genius

What's the maximum you could you write with an average pencil?

a) 10 average-length magazine articles

b) a 12-volume encyclopedia

c) the United Nations' Universal Declaration of Human Rights

d) this book

CORRECT ANSWER:

d) this book

Pencil makers claim that a pencil will write about 45,000 words. I suppose that depends on how long the words are, and how often you break the point and need to sharpen the pencil. But even with a few broken points, there should be enough lead in a pencil to write this book, provided you could do it in just one draft. (Good luck!) So how would you get pumped for the task of writing tens of thousands of words? You could do what famous novelist Ernest Hemingway did. He sharpened dozens of pencils to get himself in the mood to write. It obviously worked for him. I tried it, but I couldn't figure out what to do with all those pencils when I sat down at my computer.

Q

The first computer mouse looked like a small block of wood with a cable sticking out of it. Douglas Engelbart patented his wooden mouse in 1970. Back then, a mouse was just a novelty that you might find in a computer lab. The first popular computer mouse came with the Apple Macintosh released in 1984. When the designers were coming up with the new concept for the Mac mouse, they built various prototypes. The very first conceptual prototype was made by Dean Hovey.

What did he use to make it? A . . .

a) butter dish and a roll-on deodorant ball

b) cardboard candy box and toy truck wheels

c) juice box and a small rubber ball

d) sardine tin and a catnip ball

A Mouse in the House

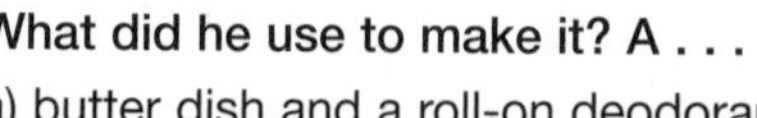

A

What did he use to make it? A . . .

a) butter dish and a roll-on deodorant ball
b) cardboard candy box and toy truck wheels
c) juice box and a small rubber ball
d) sardine tin and a catnip ball

CORRECT ANSWER:

a) butter dish and a roll-on deodorant ball

Many prototypes followed, but the concept of the deodorant ball made it into the final design. The way that Hovey and his team designed the new mouse is called rapid prototyping. Instead of waiting for parts to be manufactured, they used whatever they could find to get the ball rolling, as it were. The Mac mouse was a hit with consumers, and a mouse population boom began when other companies started releasing their own versions. These days, there's a mouse in almost every house.

First Computer Bug

Q

The Mark II Aiken relay computer at Harvard weighed about as much as a triceratops and measured about 15 metres (50 ft) by 18 metres (60 ft). This early computing behemoth was controlled by pre-punched paper tape, and could do basic math, but very slowly by today's standards. Needless to say gigantic, one-of-a-kind computers like the Mark II Aiken relay have been "extinct" for decades, yet some of the terms we use today have been around since the early days of computing.

What was the first computer bug?

a) moth

b) Trojan horse

c) virus

d) worm

First Computer Bug

What was the first computer bug?

a) moth
b) Trojan horse
c) virus
d) worm

CORRECT ANSWER:

a) moth

When the Mark II Aiken relay computer kept making adding mistakes, its operators began to hunt for the cause of the malfunction. The computer had 13,000 relays, so it could have taken a long time to find the problem. But incredibly, it only took 20 minutes to figure out what was tripping up the computer — a moth trapped between the metal contact points of Relay 70 in Panel F. They removed the moth and taped it into their logbook with the comment, "First actual case of bug being found." The amusing story of how the operators had found the bug and "debugged" the computer got around. This was in 1945, and it was probably the first time that the term "debug" was used in reference to a computer.

First Computer Programmers

If you wanted to set the ENIAC computer up in your living room, it would have to be the size of a banquet hall 30 metres (100 ft) long. The ENIAC, or Electronic Numerical Integrator And Computer, was the first large-scale, electronic, digital computer that could be reprogrammed to solve a full range of problems. Those who worked the massive machine were the world's first computer programmers. Can you believe that no one thought to honour these trailblazers for more than 50 years? By the time their work was recognized, several of them had died. So, who were these computer pioneers? Maybe you know . . .

Who were the first computer programmers?

a) female mathematicians

b) military electrical engineers

c) Phi Beta Kappa fraternity members

d) telephone operators

First Computer Programmers

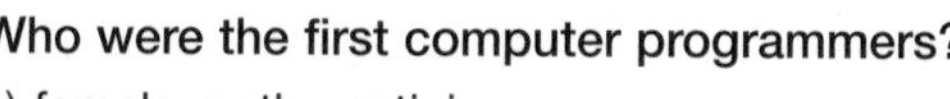

Who were the first computer programmers?

a) female mathematicians
b) military electrical engineers
c) Phi Beta Kappa fraternity members
d) telephone operators

CORRECT ANSWER:

a) female mathematicians

Most of the programming was done by six female mathematicians in the mid 1940s: Kay McNulty, Betty Jennings, Betty Snyder, Marlyn Wescoff, Fran Bilas, and Ruth Lichterman. ENIAC was a top-secret military project, and the women weren't allowed on site to watch it being built. They had to figure out how to make the computer work by memorizing drawings and diagrams. Because ENIAC had no stored programs, the women had to program every computer task by hand, by manipulating cables and switches — many, many cables and switches. Only one ENIAC was ever built. It showed that electronic circuitry worked, and the rest is computer history. Or is it herstory?

Q

There are about 100 million stars just like our sun in the Milky Way. Our sun is an average Joe, or Sol, hanging out in the suburbs of the galaxy. The sun is a middle-aged, yellow dwarf, and a bit on the cool side, but still the big guy in the solar system. If the sun wore pants, it would have to find a pair with a waist size of 4.3 million kilometres (2.7 million miles). Those are big pants to fill. How big? Have you ever wondered how massive the sun is compared not just to Earth, but to everything else in the solar system?

How much of the solar system's mass does the sun make up?

a) less than 33%

b) about 50%

c) close to 66%

d) more than 99%

Average Joe Star

How much of the solar system's mass does the sun make up?

a) less than 33%
b) about 50%
c) close to 66%
d) more than 99%

CORRECT ANSWER:
d) more than 99%

If the sun wore pants, the rest of the solar system's mass would fit into one of the pockets. Astronomers estimate that all the planets, their moons, every planetary ring, all the asteroids, meteoroids, and comets, every rock, pebble, speck of dust, and all the space junk, in the solar system add up to anywhere from 0.02% to 1% of the solar system's mass. It may not be impressive compared to the biggest stars out there, but in our solar system you can't get more massive than the sun.

The Sunshine of Our Lives

Q

What's the most important thing for sustaining life? Are you thinking food? Water? Clean air? All of those things are, without a doubt, critical to life, but the most important thing is the sun, and the energy that it beams to Earth. Not only does it power all life on the planet, but if we could collect all the sunshine that reaches us, it would be more than enough to meet the world's energy needs. Now let's take it one step farther. What if we could collect *all* of the sun's energy, not just the rays that hit the planet.

How many planet Earths could the sun power? About . . .

a) 3 million

b) 30 million

c) 3 billion

d) 30 billion

The Sunshine of Our Lives

A

How many planet Earths could the sun power?
About . . .

a) 3 million
b) 30 million
c) 3 billion
d) 30 billion

CORRECT ANSWER:
d) 30 billion

The sun's total energy output is enough to provide solar power to more than 30 billion planet Earths. We only intercept a billionth of the sun's output, and half of that gets scattered or bounced off the atmosphere. Still, the amount of solar power that gets through is nothing to sneeze at. If you took the 187 cubic kilometres (116 cubic miles) of water in Lake Erie (one of the Great Lakes), and turned it into fuel oil, and you burned it all in one second, it would produce the same amount of energy as the sunlight that strikes Earth in a day. That's a lot of solar power.

A Stellar Old Age

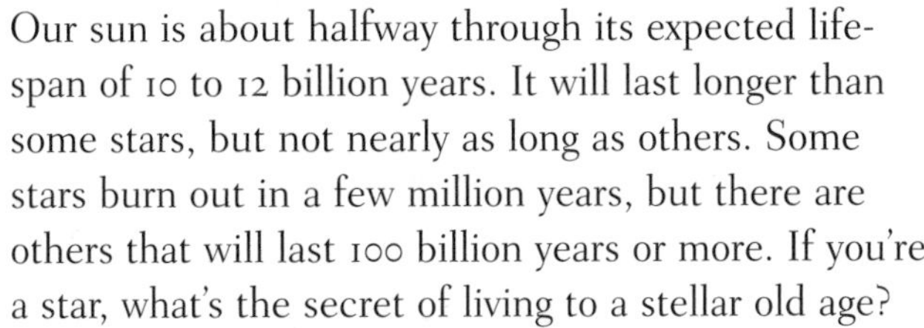

Our sun is about halfway through its expected lifespan of 10 to 12 billion years. It will last longer than some stars, but not nearly as long as others. Some stars burn out in a few million years, but there are others that will last 100 billion years or more. If you're a star, what's the secret of living to a stellar old age?

The stars that burn the longest are . . .

a) 100 times as massive as our sun

b) 10 times as massive as our sun

c) 5 times as massive as our sun

d) less than half the mass of our sun

A Stellar Old Age

The stars that burn the longest are . . .

a) 100 times as massive as our sun

b) 10 times as massive as our sun

c) 5 times as massive as our sun

d) less than half the mass of our sun

CORRECT ANSWER:

d) less than half the mass of our sun

When it comes to stellar longevity, size matters. Red dwarfs are the smallest stars, and last the longest, because they burn cooler. Burning cool means they're dim, and that makes them hard to locate, never mind study. Even the largest red dwarf known is only a tenth as bright as our sun. But just because you can't see them, doesn't mean they're not there. Some astronomers estimate that 70% of the stars in the Milky Way are red dwarfs with less than half the mass of the sun, and they should outlast Sol by at least 40 billion years. A red dwarf with a tenth of the sun's mass will shine dimly for about 100 billion years, maybe even longer. Can you even imagine living *that* long?

Q

The very first astronauts went where no man had gone before and, often, it was a one-way trip. Many of them died for our benefit, yet most of these fine, furry astronauts are forgotten. Animals were "recruited" to find out, among other things, what would happen to them (read us) in a weightless environment. More than 50 years ago, no one knew stuff like that, so instead of risking human lives, they launched a variety of animals. Animal astronauts were the first living beings in space, and the first to orbit the Earth.

Who was the first earthling in orbit?

a) Ham, a chimp

b) Laika, a dog

c) unnamed mouse

d) Yorick, a monkey

First Earthling in Orbit

Who was the first earthling in orbit?

a) Ham, a chimp
b) Laika, a dog
c) unnamed mouse
d) Yorick, a monkey

CORRECT ANSWER:
b) Laika, a dog

The first earthling in orbit was a three-year-old, six-kilogram (13 lb), stray mutt from the streets of Moscow. Laika was called Kudryavka (Little Curly) during astronaut training, which included weeks of being confined to progressively smaller boxes, and being spun in a centrifuge to simulate the G-forces of blast-off. Before her flight, she was given her astronaut name, Laika (Barker). If the mission succeeded, the little dog would make news headlines worldwide, so her name had to be easy to pronounce. She was alive and barking for the first three or four orbits (about six hours), but by the time fame came, she was dead from stress and overheating in her unshielded Sputnik 2 capsule. Laika is a hero in Russia, and her likeness is included in a monument to the cosmonauts in Star City, near Moscow. She represents the many doggy astronauts who gave their all for space science.

First Primate in Orbit

The Russian space program launched dogs because they thought that they were better than primates at being confined to small spaces for long periods of time. The American space program preferred primates because their physiology was closer to human, and they could be trained to perform tasks. They launched the first chimp astronaut, Ham, on a suborbital flight (meaning he didn't make it into orbit) in 1961. A few months after Ham returned to Earth safely, the first primate was launched into orbit.

Who was the first primate to orbit the Earth? A . . .

a) chimp, Enos

b) human, Yuri Gagarin

c) rhesus monkey, Able

d) squirrel monkey, Gordo

First Primate in Orbit

Who was the first primate to orbit the Earth? A . . .

a) chimp, Enos

b) human, Yuri Gagarin

c) rhesus monkey, Able

d) squirrel monkey, Gordo

CORRECT ANSWER:

b) human, Yuri Gagarin

The Russian cosmonaut orbited the planet once on April 12, 1961. At the end of the 108-minute mission, he ejected from his Vostok 1 spacecraft, and parachuted 1,500 metres (5,000 ft) to Earth. Gagarin's single orbit was enough to make him the first primate, and the first human, to circle the globe. He beat out Enos, an American chimp, by seven months. Enos completed two orbits, but had to settle for the titles of first non-human primate, and first chimp, in orbit. He paved the way for John Glenn, who became the first American human in orbit. He topped Enos' record by taking three spins around the planet.

First Feline on the Launch Pad

Along with the usual dogs, monkeys, chimps, and mice, many other kinds of animals have been launched into the wild blue yonder — newts, fish, frogs, tortoises, rats, rabbits, worms, spiders, insects, and cats. Cats? It does seem rather odd, given the quirky nature of felines. They're not exactly known for doing what humans want them to do, but all the first astrocat had to do was sit there and think kitty thoughts.

Which country launched the first cat astronaut?

a) Britain

b) China

c) France

d) Hungary

First Feline on the Launch Pad

Which country launched the first cat astronaut?

a) Britain
b) China
c) France
d) Hungary

CORRECT ANSWER:
c) France

Felix, a black-and-white cat from the streets of Paris, was promoted from stray to astronaut, and launched in 1963. Her brain activity was monitored with the electrodes implanted in her brain. Maybe they were planning to start an astrocat program, because it's hard to imagine how this experiment could possibly benefit human space explorers. Felix obviously had at least one of her nine lives left because she was recovered alive after her capsule parachuted 193 kilometres (120 miles) back down to Earth. A second French cat launched six days later wasn't as lucky, and joined the long list of animals who died so that humans could travel safely through space.

Q

With dogs, what you see is what you get, but sometimes we don't get what they do. Have you ever seen a dog gleefully rubbing its face and body in vile, stinky things — stuff like the droppings or rotting carcasses of other animals? We may find it quite disgusting, but the dogs rolling in it sure seem to be enjoying themselves. What's up with that? You'd have to be a dog to know for sure, but there's probably a reasonable explanation.

Why do dogs roll in stinky stuff?

a) it's their version of perfume

b) to annoy their human companions

c) to impress their friends

d) to mask their own odour

The Heady Smell of Success

Why do dogs roll in stinky stuff?

a) it's their version of perfume
b) to annoy their human companions
c) to impress their friends
d) to mask their own odour

CORRECT ANSWERS:

a), c), and d)

Give yourself a virtual point for every correct choice. Some experts think that dogs roll in stinky stuff instinctively, just like wolves, to mask their body odour. Before a hunt, wolves sometimes roll in the droppings of their prey, so that they can sneak up without their wolfy odour giving them away. That makes sense. Another idea is that the stench is an irresistible, intoxicating perfume to dogs. Could well be. Some experts even think that dogs might do it to impress their animal friends. Their animal friends might be impressed, or at the very least intrigued, but the same can't be said for their human friends.

The Faint Smell of Success

Dogs may revel in smelling like excrement or dead things, but it's not for cats. Instead of trying to smell like their chosen prey, cats try to be as odour-neutral as possible, just like the most ferocious, solitary hunters in their family tree. Like them, house cats are stealthy hunters who stalk their prey patiently, and close in slowly, until they attack with explosive force and lightning speed. There's still a wildcat inside every kitty. They may eat canned food and sleep on a pillow, but they still have the instincts and the genes of a stealthy, unsmelly hunter.

Why are cats more or less odourless? Because . . .

a) their glands produce odourless compounds

b) their saliva is a powerful cleanser

c) they bask in the sun

d) they don't eat garlic

The Faint Smell of Success

Why are cats more or less odourless? Because . . .

a) their glands produce odourless compounds

b) their saliva is a powerful cleanser

c) they bask in the sun

d) they don't eat garlic

CORRECT ANSWER:

b) their saliva is a powerful cleanser

Cats are clean and relatively unscented thanks to their cleansing saliva and countless hours of grooming, but they're not completely odourless. Aside from the obviously smelly bits, cats have scent glands at the base of the tail, on the pads of the paws, and on the temples, lips, and chin. When a cat rubs its face into you, and winds its tail around your leg, even though you can't smell it, you are being annointed with its unique, personal odours. It's one of the ways that cats mark their territory. Cat owners can't be blamed for thinking the gestures are expressions of kitty love because, in a way, they are. The cat has to love you, or at least like you, to want to mark you as its personal property.

Is it Reigning Cats or Dogs?

Pets are more popular than ever before. In most Western nations, more than half of all households have a pet. Most Canadians and Americans let their pampered cats and dogs sleep on the bed. They buy them presents, celebrate their birthdays, and may be more affectionate with their pets than with the people in their lives. Of the pet owners surveyed, 63% admitted to saying "I love you" to their pet at least once a day. Nine out of ten agreed that pets are a good source of affection. Clearly, we love our animals, but which do we love more, dogs or cats?

In most countries where more than half the households have pets . . .

a) there are more cats than dogs
b) there are more dogs than cats
c) there are more households with cats
d) there are more multiple dog households

Is it Reigning Cats or Dogs?

In most countries where more than half the households have pets . . .

a) there are more cats than dogs
b) there are more dogs than cats
c) there are more households with cats
d) there are more multiple dog households

CORRECT ANSWER:

a) there are more cats than dogs

In the United States alone, there are more than 90 million cats and close to 75 million dogs. Three quarters of dog owners have only one dog, but cat households usually have two or more. In Europe, about 47 million cats act superior to 41 million dogs. Going by the numbers, cats win in most Western nations, but there are still countries sticking, doggedly, to their canines. Dogs outnumber cats by five to two in Brazil, and in Costa Rica, where most households have a pet, by almost four to one. In Japan, there are about ten dogs for every seven cats, but that could change. Felines are getting more popular all the time. Their numbers have been rising in pet-friendly nations since the mid-1990s, and where cat numbers increase, dog numbers tend to drop.

Q

What do you get when you combine rumours of alien spaceships in Roswell, New Mexico; a whack of science fiction films; and ongoing media fascination? You get the UFO craze of the 1950s. Flying saucers captured the imagination of the world, and at least one aeronautical designer. John Frost thought that the disc's aerodynamic shape would make for the fastest, most maneuverable flying machine ever. He wanted to create the first vertical takeoff aircraft — in the form of a flying saucer. So, did his idea fly?

What happened to Frost's flying saucer?

a) Canada's Defence Department funded its development.

b) During U.S. military test flights, people mistook it for a UFO.

c) It ended up decorating the roof of a café in Roswell.

d) It flew at 500 km/h (300 mph) at an altitude of 3,000 m (10,000 ft).

What happened to Frost's flying saucer?

a) Canada's Defence Department funded its development.

b) During U.S. military test flights, people mistook it for a UFO.

c) It ended up decorating the roof of a café in Roswell.

d) It flew at 500 km/h (300 mph) at an altitude of 3,000 m (10,000 ft).

CORRECT ANSWER:

a) Canada's Defence Department funded its development.

Canada's government and Frost's employer, Avro Canada, funded his work for the first few years, and after a great sales pitch and demonstration by Frost, the U.S. military decided to invest in his dream. The saucer-shaped Avrocar was supposed to be a flying jeep that would clip along at 500 km/h (300 mph), at an altitude of 3,000 metres (10,000 ft). But $10 million* and two prototypes later, it only managed a tenth of that speed, and if it reached "altitudes" higher than 1 metre (3 ft), it became unstable. All the backers lost interest, and the top-secret flying saucer project, underway near Toronto, Canada, was grounded for good in 1961.

* If you account for inflation, $10 million is about $68 million today.

Flying Pie Tins

Q

A frisbie by any other name is a flying disc, and the very first frisbies were metal pie tins. About 125 years ago, William Russell Frisbie's bakery supplied pies to universities in the northeastern U.S., and because video games hadn't been invented yet, students flew the empty metal pie tins for fun. They were the only frisbies you could get for the next 60 years or so, until two guys in California, Warren Franscioni and Walt Morrison, came up with a revolutionary flying disc made out of plastic.

What were their flying discs called?

a) Flyin' Saucers

b) Frisbees

c) Pluto Platters

d) Martian Flyers

Flying Pie Tins

What were their flying discs called?

a) Flyin' Saucers
b) Frisbees
c) Pluto Platters
d) Martian Flyers

CORRECT ANSWER:

a) Flyin' Saucers

Franscioni and Morrison's discs were sold as Flyin' Saucers. It was a catchy name that was in sync with the UFO craze of the times, but the discs didn't exactly fly off the shelves. Morrison refined the disc by adding a slope to the outer third, called the Morrison Slope. The new and aerodynamically improved discs, called Pluto Platters, were launched, but they didn't really take off either. It was only after another design overhaul, and Wham-O®'s marketing of playing Frisbee as a new sport, that the flying disc finally became a huge hit in the late 1960s — more than 125 years after the first Frisbie pie plate was flung.

Fantastic Plastic Flying Discs

Once the flying disc caught on, there was no stopping its fans. These days there are flying disc competitions for men, women, children, seniors, people in wheelchairs, and dogs, to name a few. There are regular discs, mini discs, and the specialized discs used in disc golf, a game similar to regular golf. The first official disc golf course opened in California in 1975, but these days, there are more than 2,000 disc golf courses worldwide. All that activity keeps more than 60 flying disc manufacturers in business. Maybe you knew these things, but what else do you know?

Q

Which statement about flying discs is NOT true?

a) A Frisbee innovator's ashes were put into special edition discs.

b) A sea lion holds the world record for distance in the non-human category.

c) About 100 million Frisbees have been sold in the past 50 years.

d) The U.S. Navy built a Frisbee launcher.

Fantastic Plastic Flying Discs

Which statement about flying discs is NOT true?

a) A Frisbee innovator's ashes were put into special edition discs.

b) A sea lion holds the world record for distance in the non-human category.

c) About 100 million Frisbees have been sold in the past 50 years.

d) The U.S. Navy built a Frisbee launcher.

CORRECT ANSWER:

c) About 100 million Frisbees have been sold in the past 50 years.

That statement is NOT true. It's estimated that about 200 million Frisbees have been sold — more than the number of baseballs, footballs, and basketballs combined. The U.S. Navy, in addition to building a Frisbee launcher, spent about $400,000 to test discs in a wind tunnel for a flare launcher prototype. "Steady Ed" Headrick's ashes were pressed into discs after he was cremated. He came up with the ridges on the top of the Frisbee, called the Lines, or Rings, of Headrick, to stabilize its flight, and he's also known as the father of disc golf. A sea lion set a world record by throwing a flying disc 9.76 metres (32 ft) at a zoo in St. Louis, Missouri, in 1996. The record for the longest non-human catch, 118.8 metres (390 ft), was set two years earlier by a Californian whippet hound named Cheyenne-Ashley.

Q

What do you have in common with ants, sharks, apple trees, hummingbirds, and elephants? Here's a hint. It's simple, but more complex than anything inanimate. Another hint? It's small. An amoeba consists of only one. One what? Ah yes, a single cell. Cells are like the teeny, tiny bricks of anatomical structure. We all know they're tiny, but how tiny are they really? Maybe you've heard an old saying about angels dancing on the head of a pin? Well, what if simple cells could dance like angels?

About how many cells could crowd onto the head of a pin?

a) 10
b) 100
c) 1,000
d) 10,000

Cell Mates

About how many cells could crowd onto the head of a pin?

a) 10

b) 100

c) 1,000

d) 10,000

CORRECT ANSWER:

c) 1,000

One thousand simple cells could rip up the dance floor on the head of an average pin. A typical human cell is about 10 micrometres across — that's just ten millionths of a metre, or about one 2,500th of an inch. Can you say microscopic? Here's another way to imagine how truly tiny typical human cells are. Say your hair is of average thickness, with each strand about 100 micrometres across. Ten cells could hang out, side by side, across one of your head hairs. If you've got coarse hair, five cell buddies could join them.

Forensics 101

Criminologists use DNA to identify both victims and perpetrators. If you've watched forensics shows on TV, you've seen the investigators collecting blood samples at the gory crime scene. The DNA is extracted from the blood and analyzed at the state of the art lab in record time, and by the end of the hour, science has triumphed over crime, thanks to the DNA in the blood sample. So, have you learned anything about forensic science by watching TV shows? Let's find out.

Which blood cells are used for DNA analysis?

a) platelets

b) red blood cells

c) white blood cells

d) all of the above

Which blood cells are used for DNA analysis?

a) platelets

b) red blood cells

c) white blood cells

d) all of the above

CORRECT ANSWER:

c) white blood cells

From watching forensics shows, you might get the impression that red blood cells are used in DNA analysis. But they can't be used, and neither can other blood cells, called platelets, because these types of cells don't contain DNA. White blood cells do contain DNA, and they're the ones that are analyzed in real forensics labs.

How Human Are You?

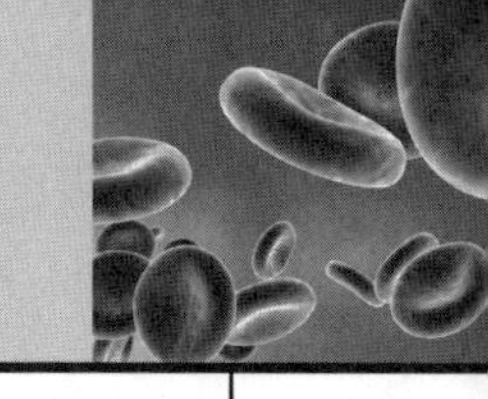

One hundred trillion. That's an estimate of how many cells make up an adult human body. Hard to imagine, isn't it? When you look in the mirror, you see a flesh-and-blood human being, not a huge collection of cells. Based on what you can see, you might assume that all your cells are true human cells. Ahhh, but are they really?

What percentage of the cells in your body are true human cells?

a) 10%
b) 30%
c) 60%
d) 90%

How Human Are You?

What percentage of the cells in your body are true human cells?

a) 10%
b) 30%
c) 60%
d) 90%

CORRECT ANSWER:
a) 10%

Only about 10 trillion of those 100 trillion cells are true human cells. Brace yourself if you're squeamish, because this means that there are close to 90 trillion bacteria crawling around in your body! Most of them make a living in your gut. Your gastrointestinal tract is sterile when you're born, but by the time you're just one month old, about 500 species of bacteria have moved in and set up permanent colonies. It's a good thing too, because without them, we wouldn't be able to digest our food.

Q

Close relatives of today's platypus were already on the scene when the dinosaurs ruled the earth. Fast-forward 110 million years to 1799, when the first Australian platypus pelt arrives in England. No one can believe it's for real. It looks like a duck's bill has been sewn onto a beaver-like body. It has to be a hoax. The claim that the bizarre-looking creature is an egg-laying mammal doesn't fly either. Most scientists ridicule the poor platypus and insist the pelt is a fake. Can you imagine how they would have reacted if they'd been told about the platypus' secret weapon?

What is the platypus' secret weapon?

a) a honk so shrill it can shatter glass

b) skunk-like scent glands under its tail

c) toxic saliva

d) venomous spurs on its ankles

The Platypus' Secret Weapon

What is the platypus' secret weapon?

a) a honk so shrill it can shatter glass

b) skunk-like scent glands under its tail

c) toxic saliva

d) venomous spurs on its ankles

CORRECT ANSWER:

d) venomous spurs on its ankles

The platypus is the most venomous mammal on the planet. Males have a spur on each ankle that can deliver enough poison to kill a small dog, or cause a human excruciating pain. All platypuses, or platypi if you prefer, are born with the spurs, but females shed them as they mature. In males, the spurs are connected to venom glands in their thighs. Aside from self-defence, it's thought that the venomous spurs may be used during territorial fights between rival males. The poison is at its most toxic during mating season. Ouch.

Hunting Like a Platypus

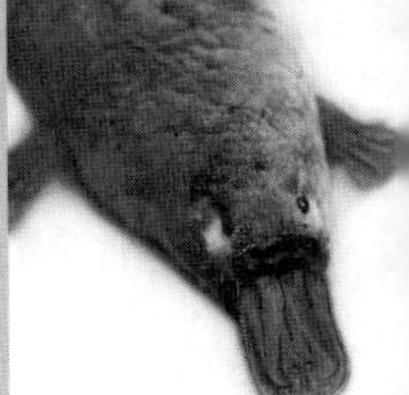

A platypus spends most of its days in the water hunting for food. Every minute or so, it returns to the surface to breathe and, if it's been successful, to eat whatever it's caught. That sounds pretty normal for a semi-aquatic mammal, but the platypus is very far from what you'd call a *normal* mammal.

Which animal hunts the most like the platypus?

a) beaver
b) hammerhead shark
c) pelican
d) sea lion

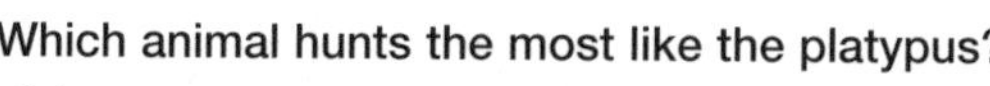

Hunting Like a Platypus

Which animal hunts the most like the platypus?

a) beaver
b) hammerhead shark
c) pelican
d) sea lion

CORRECT ANSWER:
b) hammerhead shark

What could the imposing hammerhead shark possibly have in common with the cartoonish platypus? They both have a sixth sense, powered by special organs for sensing and locating the electrical currents given off by other animals. The platypus' organs are in its velvety bill, and not only do they detect electrical fields, but they can generate them too, and detect a potential meal by the way it distorts the fields. The platypus is the only mammal that uses electrolocation. Unlike sharks, which use all of their senses to hunt, the platypus dives with its eyes, ears, and nostrils squeezed shut. It relies on electrolocation and its sensitive bill for survival. It's a unique way for a mammal to hunt, but the *modern* platypus has been around for at least 60 million years, so it must be working.

Motherhood: Platypus Style

Q

When a female platypus is ready to give birth, she goes into her breeding burrow, and plugs the hole up with leafy stuff. Then she lies down, usually on her back, and lays two, sometimes three, small, leathery eggs about the size of marbles. The eggs stick together, and to the female's fur, and she holds them against her belly with her tail. After about 10 days of this unusual form of incubation, the tiny babies break out of their shells using a "milk tooth," just like baby reptiles. What happens after that?

How does a mother platypus feed her newly hatched babies? She . . .

a) catches insects for them
b) nurses them
c) regurgitates food for them
d) teaches them how to hunt

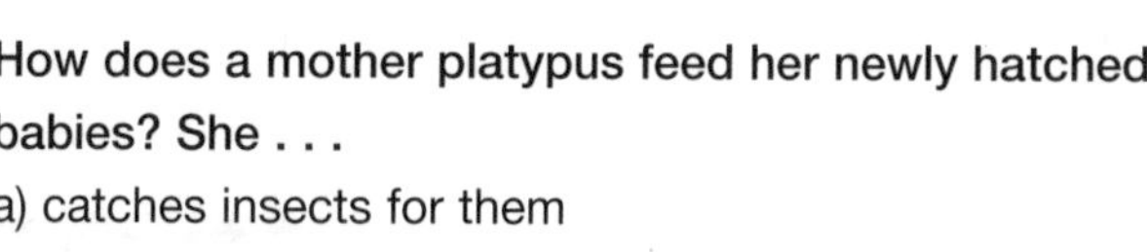

Motherhood: Platypus Style

How does a mother platypus feed her newly hatched babies? She . . .

a) catches insects for them
b) nurses them
c) regurgitates food for them
d) teaches them how to hunt

CORRECT ANSWER:
b) nurses them

Platypus females nurse their babies, despite not having any nipples! Her mammary glands secrete milk through the pores of her skin. It pools in furrows on her belly, and the tiny babies lap it up while clinging to mom's fur. Platypus babies are about the size of raisins, or soybeans, when they break out of their eggs, but after four months of mother's milk, they're about 30 centimetres (12 in) long. At that point, the juveniles leave mom's burrow and make their own way in the world. When they reach maturity, they're about 42 to 45 centimetres (16 to 18 in) long. They're not very big, but if we grew at the same rate as a platypus, an adult human would grow to be about 15 metres (49 ft) tall.

Supermales

The sun has just set on another beautiful day in paradise. It's dusk in the tropics. There's a gentle breeze, and beneath the surface of the calm turquoise waters, a small group of fish is ready to rock on the coral reef. A big, macho supermale parrotfish is getting the superstar treatment from his enthusiastic female groupies. As far as they're concerned, he's the hottest hottie on the reef, and they all want to mate with him and have his babies. They won't be disappointed. After all, he *is* a supermale. How does a guy get a job like that?

How do parrotfish become supermales?

a) dominant males fight for the position
b) females undergo a sex change
c) fertile females choose the healthiest male
d) males eat a special diet to grow bigger than the other males

Supermales

How do parrotfish become supermales?

a) dominant males fight for the position
b) females undergo a sex change
c) fertile females choose the healthiest male
d) males eat a special diet to grow bigger than the other males

CORRECT ANSWER:

b) females undergo a sex change

And no reconstructive surgery is required! Surgeonfish need not apply. You'd think that males would be the natural choice, but in a weird and wonderful parrotfish twist of nature, only females become supermales. They're all born with spare male parts. When the local supermale dies, some of the biggest females activate the male within, and one becomes the new resident supermale. The makeover is complete, both inside and out. He even changes colour. The females are attracted to his distinct new colours and markings. The supermale stage is the terminal phase for parrotfish, meaning they soon die of old age. The ability to change sex is a brilliant survival strategy. It ensures that there will always be a male to mate with the females and produce the next generation of parrotfish.

Life's a Beach

Parrotfish are always looking for something to eat. Luckily, their favourite food, the algae that grow in and around corals, are always on the menu. While they're munching on the algae, they bite off hunks of coral and grind them up with their fused teeth, which resemble a parrot's beak. Chomping on coral may sound destructive, but it's actually a win-win-win situation. The parrotfish don't go hungry, and they keep the corals healthy. The third win is a product of the parrotfish's hearty appetite. Hunks of coral go in one end, and sand comes out the other. How much sand?

How much sand can a big parrotfish make in a year?

a) 1,000 kg (2,200 lb)

b) 100 kg (220 lb)

c) 10 kg (22 lb)

d) 1 kg (2.2 lb)

Life's a Beach

How much sand can a big parrotfish make in a year?

a) 1,000 kg (2,200 lb)
b) 100 kg (220 lb)
c) 10 kg (22 lb)
d) 1 kg (2.2 lb)

CORRECT ANSWER:

a) 1,000 kg (2,200 lb)

No one has actually followed a big parrotfish around for a year and collected and weighed the sand that comes out of its rear end. But according to the best estimates and when you consider that a parrotfish spends most of its waking hours grazing on coral, it's entirely possible that it could process a whole metric tonne of sand in a year. So, if you're lucky enough to find yourself on a tropical beach, blissfully wriggling your toes in fine, white sand, remember the parrotfish and its contribution to the environment. Parrotfish are some of the unsung heroes of the reef. In an Australian study, damaged coral reefs that were grazed by parrotfish recovered, while those that were deprived of parrotfish died from an overgrowth of algae.

Better than the Best Tropical Fish Tank

The most complex, densely packed ecosystems on the planet are tropical rainforests and coral reefs. A whopping 25% of all ocean fish species, as well as hundreds of thousands, if not millions, of varieties of life call coral reefs home. The numbers are even more amazing when you consider that reefs cover less than 0.1% of the sea floor. All the world's coral reefs, put together, cover less area than Spain. They're spread out across the world in 80 geographical locations. When it comes to having the largest areas with coral reefs . . .

What are the top four countries?

a) Australia, Fiji, Belize, Mexico
b) Belize, Australia, U.S., Cuba
c) Indonesia, Australia, Philippines, France
d) Philippines, Indonesia, Papua New Guinea, Greece

Better than the Best Tropical Fish Tank

What are the top four countries?

a) Australia, Fiji, Belize, Mexico
b) Belize, Australia, U.S., Cuba
c) Indonesia, Australia, Philippines, France
d) Philippines, Indonesia, Papua New Guinea, Greece

CORRECT ANSWER:

c) Indonesia, Australia, Philippines, France

The top three countries, with half of the world's coral reefs, are Indonesia, Australia, and the Philippines. So what's France doing in the top four? You had to factor in France's many warm weather territories with impressive reefs, such as Martinique and French Polynesia. Most of the world's coral reefs are in trouble these days, due to the triple threat of climate change, pollution, and overfishing. Have you ever snorkelled or dived on a healthy coral reef? It's awesome — better than the best tropical fish tank you've ever seen, by a factor of a gazillion. If the underwater rainforest experience appeals to you, get to it as soon as you can. If conditions don't improve, the reefs will never be better than they are today.

Q

If you've ever tried rock climbing, or even just watched someone else do it, you know that it takes balance, coordination, and strength. Sometimes climbers have to hang on by their fingertips. So they must have super strong fingers, right? Have you ever wondered how many muscles there are in your fingers? If not, you will now.

How many muscles are there in your fingers?

a) 0

b) 12

c) 14

d) 28

Powerful Digits

How many muscles are there in your fingers?

a) 0
b) 12
c) 14
d) 28

CORRECT ANSWER:

a) 0

Would you believe there isn't a single muscle in any of your fingers? Your grip strength comes from muscles in your palms and forearms. Your finger bones are connected to those muscles by tendons, which pull on, and move your fingers, a lot like the strings of a marionette. In a sense, your fingers move by remote control, and they're controlled not just by tendons, but by the puppet master in your head, your brain.

A Digital Advantage

Athletes are always looking for the edge in competition, that special something, that little bit extra, that can make the difference between winning and losing. But no matter how hard they train, or how mentally prepared they are, there are some things that athletes can't learn or practice because they're either born with it, or they're not. Like toes, for example.

What would be an advantage for a sprinter?

a) big toe and second toe of equal length

b) big toe longer than second toe

c) second toe longer than big toe

d) short toes

A Digital Advantage

What would be an advantage for a sprinter?

a) big toe and second toe of equal length

b) big toe longer than second toe

c) second toe longer than big toe

d) short toes

CORRECT ANSWER:

b) big toe longer than second toe

A long, strong big toe can work to a sprinter's advantage. Leaning his or her full body weight into a robust, firmly planted big toe lets a sprinter accelerate more quickly than runners with smaller, weaker piggies. A strong big toe is also a plus in skiing. You can't cut an edge without planting your big toe. The bigger and stronger it is, the better, for some sports anyway. For non-athletic types the big piggy's edge may cut in a whole other way. According to my unofficial survey, the longer the big toe, the more quickly it pokes holes in your socks.

Significant Digits?

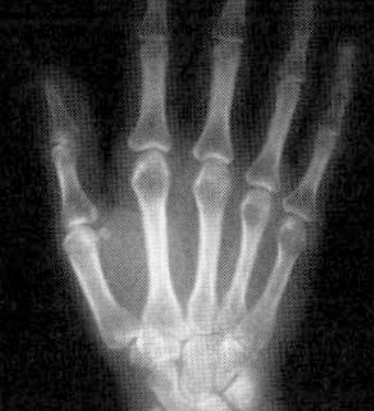

Q

Take a look at your fingers. Which is longer — your index finger, or your ring finger? If you're female, chances are that both are about the same length. If you're male, your ring finger is probably longer than your index finger. In both sexes, finger length is influenced by exposure to male hormones in the womb before birth. Finger length has been studied more in men than in women, and some researchers believe that the relative length of a male's fingers may be subtle clues to his mental and physical makeup.

Men with a significantly longer ring finger are more likely to be . . .

a) better at reading maps
b) better athletes
c) physically aggressive
d) all of the above

Significant Digits?

Men with a significantly longer ring finger are more likely to be . . .

a) better at reading maps
b) better athletes
c) physically aggressive
d) all of the above

CORRECT ANSWER:
d) all of the above

On average, men whose ring fingers were a lot longer than their index fingers were found to be all of the above, in various studies. How seriously should we take this stuff? At this point, probably not too seriously. If the studies are to be believed, males with ring and index fingers closer to the same length are more prone to depression, but women with the same finger ratio are not. Some studies have found that men and women with longer ring fingers are more likely to be gay, but in other studies, no correlation between finger length and sexual orientation was found. My vote for the most amusing finding goes to this little gem: women with longer ring fingers slammed the phone down harder when they were frustrated. I think we need a study to determine the relative finger lengths of the people who come up with studies like that!

Q

Hot on the trail of working germs, The Germinator, a.k.a. microbiologist Dr. Chuck Gerba, with his tireless team, collected 7,000 samples from a variety of offices. One of the items that they tested were coffee mugs. Everyone's mugs were sampled, from the presidents and higher ups, down to the lowliest workers. There turned out to be more in the mugs than just steaming hot coffee, and it wasn't sweet or white.

Whose mugs contained the most fecal bacteria?

a) data processors

b) middle managers

c) technical support

d) top executives

A Steaming Hot Cup of . . .

Whose mugs contained the most fecal bacteria?

a) data processors

b) middle managers

c) technical support

d) top executives

CORRECT ANSWER:

d) top executives

The higher up in the company they were, the more fecal bacteria in their coffee mugs. It seems counterintuitive to find the poopy germs associated with toilets in the mugs of the most pampered workers in the corporate world. Dr. Gerba thinks it's because the top executives had people to clean their mugs for them. The ones doing the washing up, and serving coffee to the boss in a "fresh" mug, were using sponges or dishrags contaminated with fecal bacteria. They washed the mugs, but not the dishrags, which the germs were busy populating. Most regular workers weren't as vigilant about washing their own mugs, and so didn't smear germs into them with the dishrag as often. Finally, a legitimate excuse for not cleaning your dirty mug at work.

The Top 10 Most Contaminated Things in Your Office

The Germinator and his team sampled various surfaces in offices including desktops, phones, computer mice, keyboards, microwave door handles, elevator buttons, photocopier buttons, photocopier surfaces, toilet seats, fax machines, refrigerator handles, and water fountain handles. They were hunting for pathogens — the bad bacteria, such as E. coli, pneumonia, and salmonella, which can make you really sick.

Where did they find the most germs?

a) elevator button

b) phone receivers

c) photocopier start button

d) toilet seats

The Top 10 Most Contaminated Things in Your Office

Where did they find the most germs?

a) elevator button
b) phone receivers
c) photocopier start button
d) toilet seats

CORRECT ANSWER:

b) phone receivers

Dr. Gerba says that there are more bacteria and viruses in office cubicles than in public restrooms. The average phone receiver had 500 times as many bacteria as the average toilet seat, which was close to the bottom of the list. The least contaminated items were doorknobs and light switches, probably because they weren't touched nearly as often as phones or computers.

The 10 Most Contaminated Things in the Average Office

1. phone
2. desktop
3. water fountain handle
4. microwave door handle
5. computer keyboard and mouse
6. inside the desks of female workers
7. palm pilots
8. photocopier buttons
9. fax machine buttons
10. sink in the coffee prep area

Hand Island

People do more than just work at their desks; they live at them too. For eight or more hours a day, they eat meals and snacks, pay bills, talk on the phone, answer personal email, and so on. While they're doing all of the above, and maybe even getting some work done, most people have a spot on the desk where they typically park their hand. The Germinator knew about the handy hotspot, and tested it.

How many germs did that area contain on average? About . . .

a) as many as a toilet
b) 4 times as many as a toilet
c) 100 times as many as a toilet
d) 400 times as many as a toilet

Hand Island

How many germs did that area contain on average? About . . .

a) as many as a toilet
b) 4 times as many as a toilet
c) 100 times as many as a toilet
d) 400 times as many as a toilet

CORRECT ANSWER:
d) 400 times as many as a toilet

You might as well call the area "Hand Island," because it could be home to 10 million or so bacteria. Say you used the toilet at work, and didn't wash your hands. Would you pick your nose, rub your eyes, or lick your fingers? Maybe not. Guess what? Hand Island is probably more contaminated than the office toilet. If the idea of sharing your desk with millions of germs doesn't appeal to you, The Germinator recommends annihilating them with a disinfectant wipe once a day. Hand Island's population will be cut by about 99.9%.

Q

Imagine it: you're on a TV game show, and the smarmy host has just told you that you've won a billion dollars. Before you can shut your gaping mouth he continues, "Buuut . . . there IS . . . a catch. You . . . can only keep . . . as much money . . . as you . . . can count!" You're ready for the challenge! Let the counting begin! "Annnd . . ." the obnoxious host continues, with even more dramatic pauses, "it's . . . all . . . in . . . one . . . dollar . . . bills!" The studio audience goes wild when you give the cameras a huge grin and two thumbs-up. You're ready to count your way to a billion dollars! It takes one second to count each dollar.

About long does it take you to count a billion dollars?

a) 32 hours

b) 32 days

c) 32 months

d) 32 years

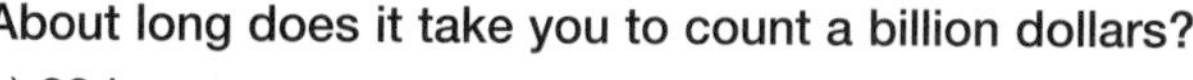

Let the Counting Begin!

About long does it take you to count a billion dollars?

a) 32 hours

b) 32 days

c) 32 months

d) 32 years

CORRECT ANSWER:

d) 32 years

Sadly, the game show would be cancelled long before you could count your billion. If you do the math (not counting leap years), it actually works out to 31 years, 259 days, 1 hour, 46 minutes, and 40 seconds . . . almost 32 years without a bathroom break or a wink of sleep! And that's counting an American billion (a thousand million). You'd have to live almost 3,200 years to count a British billion (a million million)! These days, a billion usually refers to the American billion, even in Britain.

Birdy Body Count

There are billions and billions of birds on the planet. They're divided into just under 10,000 species of which 70% are declining in numbers, and of those, 10% are winging toward extinction. But one species is experiencing a spectacular population boom, and may turn out to be the most successful bird species ever. There are more of them today than any other kind of avian. So why are they booming while other bird species are going bust? Because we keep producing them!

How many chickens are there on the planet?

a) as many as there are stars in the Milky Way galaxy
b) four times as many as the human population
c) more than all the rest of the birds in the world combined
d) one gigabyte, if each chicken is one byte

Birdy Body Count

How many chickens are there on the planet?

a) as many as there are stars in the Milky Way galaxy
b) four times as many as the human population
c) more than all the rest of the birds in the world combined
d) one gigabyte, if each chicken is one byte

CORRECT ANSWER:

b) four times as many as the human population

There were more than 24 billion chickens clucking on the planet in 2007. That's about four chickens for every man, woman, and child. Unless a bird flu or other disease has decimated their numbers, there are probably even more of them now. By 2100, a quarter of all bird species could be extinct. What are the odds that chickens will be among them? Even if they did get into trouble, is there any doubt that heroic efforts would be made to save the chickens, and bring their population back up into the billions as quickly as possible? I'd count on that.

It's What's Inside that Counts

Ever heard the saying, "It's what's inside that counts"? When it comes to your body, it couldn't be more true. We're totally unaware of the life and times of the trillions of cells that make up our bodies, even though our lives depend on their proper functioning. A healthy body stays in balance by replacing cells at the same rate as they're lost. Cells are constantly dying, and being born. Take red blood cells: they develop in your bone marrow, and pop out into your bloodstream every day. How many of them in a day? Excellent question!

How many red blood cells does your body make in 24 hours?

a) 1 billion
b) 2 billion
c) 20 billion
d) 200 billion

It's What's Inside that Counts

How many red blood cells does your body make in 24 hours?

a) 1 billion

b) 2 billion

c) 20 billion

d) 200 billion

CORRECT ANSWER:

d) 200 billion

In fact, every day, your body makes *more* than 200 billion red blood cells to replace the same number of worn-out cells. Each and every second, about two million red blood cells are born. If you've spent a minute reading this far, in that 60 seconds, 120 million new red cells have popped out of your bone marrow and joined your bloodstream. If you read for another 10 minutes, you'll have 1.2 billion more fresh, young, red cells in action. The ones that they've replaced have been removed from circulation, and are being broken down right now. The iron inside them is being recycled and reused to make — what else? — more red blood cells. Bloody amazing!

These words are all FAT, meaning phat, cool, bad, awesome, brilliant, or neat. They have to do with things on previous pages, or pages yet to come, and they all start with F.

What do you think they mean?

Ferrule

a) dominant male ferret with a large territory
b) metal band that holds the eraser to the end of a pencil
c) small funnel used by perfume chemists
d) tool for testing the iron content of ore

Fomite

a) gel for sopping up oil spills
b) inanimate object contaminated with germs
c) plastic laminate used in kitchens and bathrooms
d) styrofoam packing pellets

Furcula

a) bat's roost
b) Dracula's cat
c) forked bone found in birds and therapod dinosaurs
d) root of animal hair follicles

Ferrule

CORRECT ANSWER:

b) metal band that holds the eraser to the end of a pencil

One of the final parts of pencil-making is attaching the ferrule and the eraser to the end of the pencil. The word refers not just to the metal band on the end of a pencil, but to other metal rings, caps, or sleeves, as well.

Fomite

CORRECT ANSWER:

b) inanimate object contaminated with germs

Any inanimate object or thing that holds or absorbs germs or parasites and transfers them from one person to the next is called a fomite. For example, a doorknob covered in flu virus is a fomite, as is a comb infested with head lice.

Furcula

CORRECT ANSWER:

c) forked bone found in birds and therapod dinosaurs

You've probably seen a chicken's furcula. Maybe you've even wished on one. Furcula means "little fork" in Latin, and it's a wishbone by any other name. It strengthens birds' chest muscles for flying.

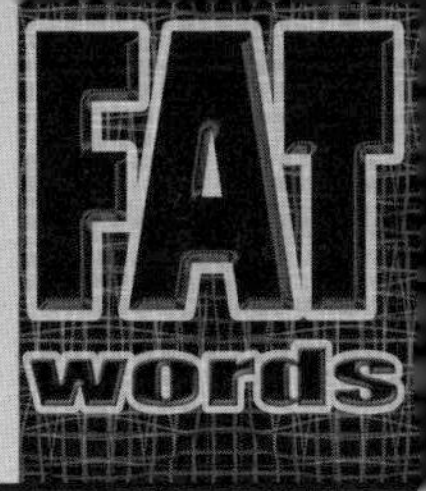

These terms are FAT too, and they also have something to do with things on previous pages, or pages yet to come. How well do you know your ABCs?

What do you think they mean?

Astronomical Unit

a) currency for space tourists
b) distance from the Earth to the sun
c) measure of a star's mass
d) team of four astronomers

Borborygmus

a) short-sightedness
b) sounds your belly makes when you're hungry
c) three-dimensional geometric shape
d) vortex of flushing toilet water

Chalaza

a) cheese-filled fried flatbread
b) egg yolk's anchor
c) hammock chair
d) hollow gourd used as a percussion instrument

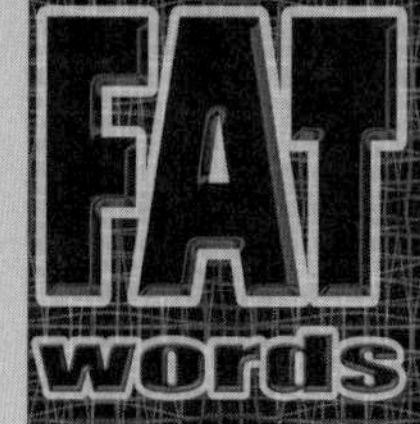

Astronomical Unit

CORRECT ANSWER:

b) distance from the Earth to the sun

If you want to be precise, the generally accepted value of an astronomical unit, or AU, is 149,597,870,691 kilometres, plus or minus 30 metres. Or you could just think of it as being about 150 million kilometres, or 93 million miles — the distance from the Earth to the sun.

Borborygmus

CORRECT ANSWER:

b) sounds your belly makes when you're hungry

It's that gurgling, burbling, rumbling sound your belly makes when you're hungry. The sound comes from the movement of gases and liquids making their way through the gut. The word comes from the ancient Greeks, who were trying to imitate gut sounds with the word.

Chalaza

CORRECT ANSWER:

b) egg yolk's anchor

The plural of chalaza is chalazae, and they're the structural fibres that hold an egg yolk in place. If you've ever cracked open a raw egg and seen a thick, white, ropelike strand, you've seen the chalazae. If you want to know how fresh an egg is, check out the chalazae: the thicker the strand, the fresher the egg.

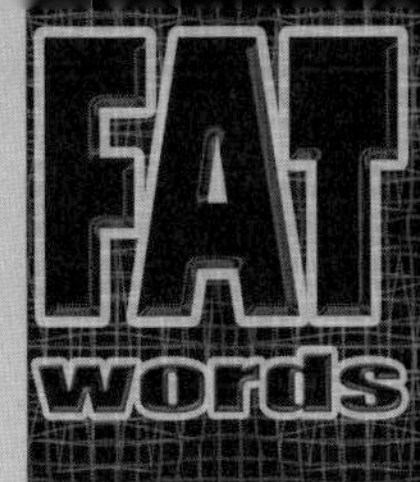

These words are also FAT, and they all start with T.

What do you think they mean?

Tittle
a) dot over the letter i
b) mammary gland of whales
c) snicker
d) unit of fluid measurement

Tribology
a) oral record of tribal history
b) psychology of same culture bonding
c) science of interacting surfaces in relative motion
d) study of the history of newspapers

Trophallaxis
a) angle of the earth's axis
b) food regurgitation by one animal for another
c) insect mating act
d) laxative

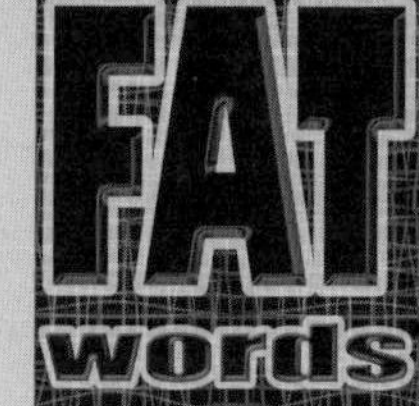

Tittle

CORRECT ANSWER:

a) dot over the letter i

In English, the letter i is always dotted with a tittle, but the dot over the i is not the only tittle. The marks made over a letter, such as accents, circumflexes, umlauts, or under a letter, such as the cedilla used under the letter ç, are also tittles.

Tribology

CORRECT ANSWER:

c) science of interacting surfaces in relative motion

Tribology is the study of friction, wear, lubrication, and design of bearings. A bearing is the part of a machine that bears friction, especially between the rotating shaft and its housing.

Trophallaxis

CORRECT ANSWER:

b) food regurgitation by one animal for another

It sounds gross, but for wolf pups it's like fast food delivery. Wolves, and other members of the dog family, gobble up a heaping helping of the kill, trot it home, and then regurgitate it for the pups to eat. Some birds, social bees, and ants also practice trophallaxis to share food.

Q

They don't have hearts, brains, or real eyes, but that hasn't stopped them from becoming some of most successful species on the planet. They're also among the world's oldest life forms, and evolved 260 million years before sharks, 400 million years before dinosaurs, and about 650 million years before the first humans. They can regenerate lost body parts, and when food is scarce, they can shrink in size to conserve energy. Pretty amazing, you must admit, but the most remarkable thing about them may be that they can sting you from beyond the grave. Ouch! What is this mysterious life form?

What can still sting you long after it's dead?

a) cone snail

b) jellyfish

c) scorpion

d) stinging nettle

Dead Stingers

What can still sting you long after it's dead?

a) cone snail

b) jellyfish

c) scorpion

d) stinging nettle

CORRECT ANSWER:

b) jellyfish

Of the 200 or so species of jellyfish, about 70 sting humans. Even though their stings can be painful, the vast majority won't kill you. The Australian Box Jelly is a terrifying exception. Its venom is more poisonous than a King Cobra's, and can kill a human in minutes! They can be very bad news if you happen to run into one, but there's good news about jellyfish too. They aren't threatened by extinction, they're low in calories and fat, and rich in nutrients. Jellyfish sushi, anyone? Dried, they're a popular snack in some Asian countries, especially Japan. And one last tidbit: if you change one letter in "snack," you get "smack," which is what a group of sea jellies is called.

Jiggly Food Product

It jiggles, wiggles, and sparkles, and if you believe the advertising, everybody loves it. This colourful treat comes in a wide range of mostly fruity flavours, and is semi-transparent. It's easy to digest, and literally melts in your mouth because your body temperature is higher than its melting point. We're talking gelatin. Maybe you're a fan of the jiggly treat. If not, you've probably eaten it at some time or other, and at the very least, you've been in the same room with it. But what do you really know about it?

What is gelatin made of?

a) a yam-like tuber
b) animal skins and bones
c) fruit peels
d) seaweed

Jiggly Food Product

What is gelatin made of?

a) a yam-like tuber
b) animal skins and bones
c) fruit peels
d) seaweed

CORRECT ANSWER:

b) animal skins and bones

Alginate, pectin, agar, and carageenan are some of the gelling agents from plant sources, but gelatin is made from the animal parts that nobody wants. The bones and skins of various animals, including pigs, cows, fish, and in China, donkeys, are boiled down to extract collagen, the stuff that puts the jiggle into jelly. I can hear the vegetarians, and maybe even some squeamish meat eaters, shuddering right about now. OK, it's not pretty. But on the other hand, if you consume meat anyway, and provided the products are safe to eat, what's so awful about using up all the odds and ends of the animals we kill for food? If you eat regular hot dogs, you're already doing that anyway.

The Goldilocks Zone

Certain words or phrases sound so intriguing, they're bound to fire up your imagination. Rather than just looking them up, it can be fun to let your creativity run free as you come up with possible answers to what a word or phrase might mean. Have you ever heard of the Goldilocks Zone? If you haven't heard of it, where or what do you think it could be? I've come up with four choices for you, but before you answer, feel free to come up with some possibilities of your own.

What is the Goldilocks Zone?

a) a solar system's sweet spot

b) manufacturing lingo for mattresses in the medium-firm range

c) that warm, fuzzy, sleepy feeling after eating a big meal

d) the depth at which gold is usually found underground

The Goldilocks Zone

What is the Goldilocks Zone?

a) a solar system's sweet spot
b) manufacturing lingo for mattresses in the medium-firm range
c) that warm, fuzzy, sleepy feeling after eating a big meal
d) the depth at which gold is usually found underground

CORRECT ANSWER:

a) a solar system's sweet spot

The Goldilocks Zone refers to a solar system's habitable zone — the area that's just the right distance from a star for life to evolve on an orbiting planet. It's not too close, and not too far, so it's not too hot, and not too cold, and liquid water can exist on the surface. How big a sun's habitable zone is depends on how hot it is. Our sun's Goldilocks Zone extends from our orbit to Mars' orbit, and possibly beyond, maybe even to some of Jupiter's hard, rocky moons.

Tastes Like Chicken

Would it occur to you grab a bowl of mealworms for a snack? Probably not, unless you're a Klingon and grew up eating them. Most of us prefer whatever we grew up eating, and avoid foods that are different, or exotic. Whatever our food preferences, we all eat insects without knowing it, but we're talking entomophagy here — deliberate insect eating. Fewer people worldwide are eating insects, thanks to the spread of Western squeamishness, and maybe that's not such a good thing. The bugs munching on food crops are often more nutritious than whatever they're eating. Even if they are nutritious, insects can't possibly taste good, can they? The people who eat them think otherwise . . .

Which bug tastes like chicken?

a) bee

b) cicada

c) locust

d) witchety grub

Tastes Like Chicken

Which bug tastes like chicken?

a) bee
b) cicada
c) locust
d) witchety grub

CORRECT ANSWER:

b) cicada

The taste of roasted, fried, or boiled cicadas has been compared to chicken, french fries, or popcorn. Cicada sushi tastes like raw potatoes with a hint of avocado, or maybe clam, depending on who you ask. Other tasty bugs on the edible menu include locusts, which are similar to fried shrimp, and witchety grubs, which are big, succulent moth larvae, or caterpillars. They were the most important insect food in the Australian desert, where they were enjoyed by Aborigines. Cooked in hot ashes, witchety grubs, apparently, taste like almonds. Ten of them provide a day's worth of calories, protein, and fat. Despite the nutritional punch some bugs pack, most of us consider them to be pests, not food items.

Q

Bast had the body of a shapely woman and the head of a domestic cat, but that only enhanced her appeal as one of the most popular goddesses in ancient Egypt. Cats were sacred to her, and thousands of them lived a life of luxury in her beautiful red granite temple. When they died, they were mummified. Soon enough, cat lovers took to mummifying their pets too. Some pampered pussies were buried with mummified rodents to enjoy for all eternity, but for most of them it only seemed like eternity. The peaceful afterlife of some 10,000 cats was disturbed in 1888, when a Nile farmer dug into one of the vaults in a cat cemetery, or necropolis. The cat mummies were stacked as far, and as high, as the eye could see.

What happened to these stacks of cat mummies? They . . .

a) fueled the Orient Express

b) were processed and ground up for fertilizer

c) were taken to the Cairo Museum

d) were used as ballast in ocean-going vessels

Where Have All the Cat Mummies Gone?

What happened to these stacks of cat mummies? They . . .

a) fueled the Orient Express

b) were processed and ground up for fertilizer

c) were taken to the Cairo Museum

d) were used as ballast in ocean-going vessels

CORRECT ANSWER:

b) were processed and ground up for fertilizer

Mummies were used for all kinds of things, including fuel for trains, and even ships' ballast, but the cat mummies dug up by the Nile farmer in what is now Tell Basta were processed into 17 tonnes (19 tons) of fertilizer. It was a bargain at just £4 a tonne, and was plowed into British soil. Mummies were a cheap and plentiful commodity more than 100 years ago, but these days, they're priceless. The ones that escaped being processed into mummy "products" have given us a glimpse into the cat's mysterious past.

Egyptian Cat Fancy

Ancient Egyptian cat lovers mummified millions of cats. Over thousands of years, more and more cemeteries were built to accommodate the felines, but they couldn't be built fast enough. Eventually, there was standing room only for the eternal kitties. Three hundred thousand crowded cat mummies were excavated from the necropolis in Beni-Hassan alone, but by the time scientists became interested in studying them to learn about the domesticated cat's evolution, they were fairly rare. Luckily, there were enough "surviving" cat mummies to uncover the mystery of which cats the Egyptians revered and loved.

What kinds of cats did the ancient Egyptians mummify?

a) Abyssinian cats
b) Mau cats
c) wildcats
d) all of the above

Egyptian Cat Fancy

What kinds of cats did the ancient Egyptians mummify?

a) Abyssinian cats
b) Mau cats
c) wildcats
d) all of the above

CORRECT ANSWER:

d) all of the above

Based on the shape of the skull, the ancient Egyptians kept felines that were dead ringers for today's Abyssinian and Mau cats. Unlike us, the ancient Egyptians didn't recognize breeds. They called all cats, whether wild or domesticated, Mau or Miu. They mummified lots of wildcats, and several of the same species are still found in the area today. Would some of these wildcats prove to be the queen mothers of all modern kitties?

The Queen Mothers of All Kitties

Q

A fluffy, flat-faced Persian doesn't look much like a sleek, pointy-nosed Siamese, or a hairless Rex, or a Manx with no tail. No one would blame you for thinking that they all descended from different wildcats, but the truth is just the opposite. The scientists who tested the DNA of close to 1,000 cats, and their wild relatives, think that all domestic cats descended from a handful of female wildcats who lived about 130,000 years ago. Once the genetic identity of the "Eve" cats was discovered, the mystery of their origins was revealed too.

Where did the "Eve" cats live? In the . . .

a) Fertile Crescent

b) Golden Horseshoe

c) Golden Triangle

d) Valley of the Queens

The Queen Mothers of All Kitties

Where did the "Eve" cats live? In the . . .

a) Fertile Crescent

b) Golden Horseshoe

c) Golden Triangle

d) Valley of the Queens

CORRECT ANSWER:

a) Fertile Crescent

The Fertile Crescent, in the Middle East, is considered to be the cradle of human civilization, and it was the cat's cradle too. Domestic cats descended from the most-mummified cats in Egypt, the Near Eastern Wildcat, which still lives in the Middle East. No one knows how wildcats behaved for the first 100,000 years they associated with us, but it seems likely that they finally committed to domestication between 10 to 12 thousand years ago. That's when humans took up farming in the Fertile Crescent and, from the cat's point of view, when the all-you-can-eat rodent buffet was invented. Was this the point at which cats decided that the 24/7 free lunch was worth putting up with us, maybe even being nice to us? We'll never know, but we do know that being nice to us was key to making the domestic cat the most successful feline species on the planet today.

The Cougar on the Couch

Q

Although they'd never admit it, domestic cats owe their success to having evolved alongside humans. We fell in the love with the cute, friendly ones, and thanks to our devotion, and a lack of birth control, they multiplied and conquered the pet world. Cats may owe their success to cat lovers, but cat lovers may owe some aspects of good health to their cats. Animal companions can lower our blood pressure and/or cholesterol levels, help us recover more quickly from heart attacks, and they might even help us live longer. We live with them, love them, and treat them like family members, but how much do we really know about the little cougar on the couch?

Which statement is NOT true?

a) All cats lack a sweet tooth.

b) Antarctica is the only continent with no native cats.

c) Cats can hear higher tones than dogs.

d) Most adult cats are lactose intolerant.

Which statement is NOT true?

a) All cats lack a sweet tooth.

b) Antarctica is the only continent with no native cats.

c) Cats can hear higher tones than dogs.

d) Most adult cats are lactose intolerant.

CORRECT ANSWER:

b) Antarctica is the only continent with no native cats.

Antarctica is *not* the only continent where no native cats developed. They didn't develop in Australia either. Humans took them to the land down under, and inevitably, some kitties went feral, and started breeding in the wild. Without any native predators, or competition, these feral cats are a serious menace to Australia's native species. They may be good for us, but they're not always good for the environment. Early on in the evolution of the cat family, one of two genes for tasting sweet became inactive, and as a result, no cat, big or small, has a sweet tooth. Equally surprising is that most adult cats are lactose intolerant, which means they can't digest milk. Cats can actually hear higher tones than dogs, but don't bother trying to invent a cat whistle. Cats would never respond to it.

Q

Dogs are a lot like us. They're social animals who like to live in a group and interact with its other members, and they have a social code of conduct for keeping things harmonious. Among humans, proper etiquette depends on your culture, but among dogs, things are a lot simpler. There's only one unwritten book of doggy etiquette. You can train a dog to shake a paw on command, but when two dogs meet, they're going to go by the book, and perform the not-so-secret handshake that all polite dogs know.

What is the doggy version of a handshake?

a) butt sniffing

b) ear sniffing

c) face sniffing

d) mouth sniffing

The Doggy Handshake

What is the doggy version of a handshake?

a) butt sniffing

b) ear sniffing

c) face sniffing

d) mouth sniffing

CORRECT ANSWER:

a) butt sniffing

It may not be our idea of etiquette, but it works for dogs. Ear, face, and mouth sniffing can be part of a doggy greeting, but the butt sniff is the official handshake. Each dog has an individual "scentprint" that come from glands at the base of the tail and around the anus, and wagging the tail wafts their personal aromas into the air. To make sense of those scents, dogs have a built-in decoder, a.k.a. a Jacobson's organ. Two tiny openings in the roof of the mouth, between the canine teeth, funnel the aroma to their scent decoder and brain. With a sniff or two, they know all they need to know about the other dog — whether it's male or female, how old it is, its mood, and probably even the contents of the last meal in the pipe. Too much information? Not if you're a dog, and you were born wired for smell-a-vision.

The Wolf Within

Q

When you look at toy poodles, chihuahuas, or dachshunds, it's hard to imagine that there's a wolf within, but there is, coiled up inside every dog's DNA. Modern wolves, and dogs, descended from ancestral wolves, and they're still very closely related. As one top dog expert put it, there's less difference between dogs and wolves than between frogs in a pond. So, physically they're the same, but what about their behaviour? To find out, a Hungarian experiment presented dogs and socialized wolves who'd been raised as pets with an unsolvable problem: a container of food that they couldn't reach.

What did they do?

a) Both the dogs and the wolves asked the experimenter for help.

b) Neither asked for help.

c) Only the dogs asked for help.

d) Only the wolves asked for help.

What did they do?

a) Both the dogs and the wolves asked the experimenter for help.
b) Neither asked for help.
c) Only the dogs asked for help.
d) Only the wolves asked for help.

CORRECT ANSWER:

c) Only the dogs asked for help.

Once they figured out they couldn't reach the food, the dogs wasted no time making eye contact with the experimenter and asking for help. The wolves, despite being socialized, avoided eye contact, and carried on alone. Wolves refuse to look at faces, but dogs do it all the time. Could they have learned it from us? At some point in the past — most experts think about 15,000 years ago — the dog brain developed human-friendly communications skills. But long before that, maybe even as long as 400,000 years ago, the dog's ancestors were already hanging around near our ancestors. As we evolved, our canine companions evolved along with us, into the astounding variety of shapes, colours, and sizes that we prefer. Today, there are more than 400 known dog breeds. They get my vote for the world's most successful canine species. There may be as many as half a billion dogs in the world today.

The Brain Wagging the Dog's Tail

Q

Dogs don't walk their talk, they wag it. Their tails act out how they're feeling at any given moment. To learn about the connections between a dog's brain and emotions, and its tail, Italian researchers set up an experiment with 15 male and 15 female dogs. Each dog was put in a covered cage with a small window that it could see through, and its tail was filmed wagging in response to its owner, an unfamiliar person, a male cat in a cage, and a large, dominant male dog in a cage.

How did the dogs wag their tails in response to their owners?

a) all over the place
b) mostly to the right
c) mostly to the left
d) it depended on the dog

The Brain Wagging the Dog's Tail

How did the dogs wag their tails in response to their owners?

a) all over the place
b) mostly to the right
c) mostly to the left
d) it depended on the dog

CORRECT ANSWER:

b) mostly to the right

Dog owners got the most enthusiastic response from their dogs, who wagged vigorously, with big swings toward the right side of the rump. In response to the unfamiliar person, they wagged moderately, but still to the right. Even the cat got a bit of a wag to the right, but when they saw the big, intimidating dog, their tails did a 180, and switched to wagging to the left. Dogs have two brain hemispheres. The left lobe is more about being happy and relaxed, and controls the right side of the body. When a dog's tail is wagging to the right, it's a happy dog. The right side of the brain is more about negative emotions, and it controls the left side of the body. When the subjects felt threatened by the big dog, their tails wagged to the left, the side controlled by the more fearful right hemisphere. Dogs also wagged to the left when they were alone. They just want to be wherever you are.

They Just Want Whatever You're Having

I once heard a comic joking that whenever he took the garbage out, his dog started whimpering and barking, as if to say, "What are you crazy?! Throwing out all that good food! I could eat that stuff!" At the time I had a refined pooch who, thinking he was human, ignored garbage, so I didn't get the joke. But then along came dog number two, who, just like the comic's dog, gets frantic whenever food waste or compost ingredients leave the kitchen. He lives to eat, and will eat just about anything, or try to anyway. Unfortunately, dogs don't know what's good for them. Do you?

Which foods are toxic to dogs?

a) chocolate, garlic, and onions

b) grapes and raisins

c) all of the above

d) none of the above

They Just Want Whatever You're Having

Which foods are toxic to dogs?

a) chocolate, garlic, and onions
b) grapes and raisins
c) all of the above
d) none of the above

CORRECT ANSWER:

c) all of the above

No one knows why, but grapes and raisins can cause kidney failure in dogs. Onions and garlic are good for people, but can make pets ill. A chemical in raw, cooked, or dehydrated onions causes anemia in both dogs and cats. Adding garlic to a pet's diet is often cited as a natural way to repel fleas, but it contains the same toxic chemical as onions, so it's not the best bet for flea control. Another food that some people think is safe for pets, but isn't, is chocolate. All chocolate is toxic to pets, even milk chocolate, and the darker it is, the more toxic. No amount is safe. As little as half a square of baking chocolate can be lethal to a tiny dog, and only 7 squares can kill a big one. That's on average. Some dogs are more, and some are less, sensitive. So, if your dog is hounding you for some of the chocolate that you're eating, just say no.

Q

You are unique. No one else is exactly like you — unless you have an identical twin. But if you're a singleton, your genetic fingerprint distinguishes you from every other person on the planet. People come in a wide range of shapes, sizes, and colours, and have wildly individual personalities, talents, and abilities. Obviously, at least some of those differences must be reflected in our genetic material, but how big a difference can there be, really?

Genetic material between humans differs by about how much?

a) 1%

b) 2%

c) 12%

d) 21%

Your Genetic Fingerprint

Genetic material between humans differs by about how much?

a) 1%
b) 2%
c) 12%
d) 21%

CORRECT ANSWER:
a) 1%

You're one of a kind, but the difference between you and everyone else is less than one percent. The other 99% is the same as everyone else's genetic material. That's because we're all humans and, just like every other creature, we have a genome that's unique to our species. The human genome is made of DNA in the form of 23 pairs of chromosomes that carry our genes. When it comes to our genes, humans are like so many peas in a pod.

Unravelling Your DNA

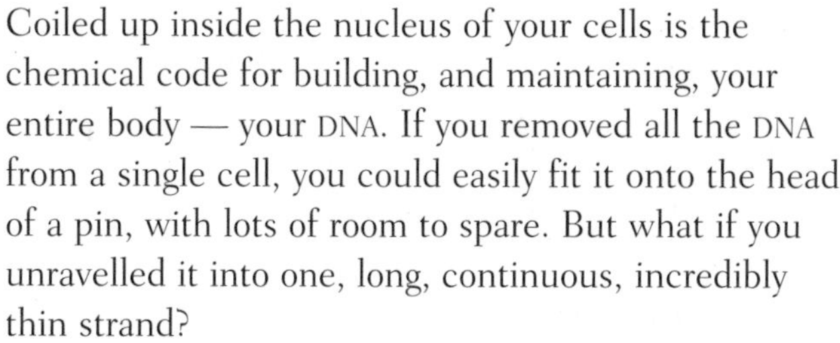

Coiled up inside the nucleus of your cells is the chemical code for building, and maintaining, your entire body — your DNA. If you removed all the DNA from a single cell, you could easily fit it onto the head of a pin, with lots of room to spare. But what if you unravelled it into one, long, continuous, incredibly thin strand?

About far would it stretch?

a) circumference of a 33-cm (13 in) pizza
b) distance from Toronto, Canada, to Ulan Bator, Mongolia
c) height of hockey great Wayne Gretzky
d) length of an Olympic-sized swimming pool

Unravelling Your DNA

About far would it stretch?

a) circumference of a 33-cm (13 in) pizza
b) distance from Toronto, Canada, to Ulan Bator, Mongolia
c) height of hockey great Wayne Gretzky
d) length of an Olympic-sized swimming pool

CORRECT ANSWER:

c) the height of hockey great Wayne Gretzky

Wayne Gretzky is 180 centimetres (6 ft) tall, about as tall as the human genome is long. Regardless of how different you may be from this famous hockey player, the length of your genome is the same. So the next time someone is annoying you by telling you that his or her whatever is bigger or better than yours, you can shoot back that your genome is as long as Wayne Gretzky's (or any other famous athlete, scientist, or artist of your choice). So there!

The Map of Your Genes

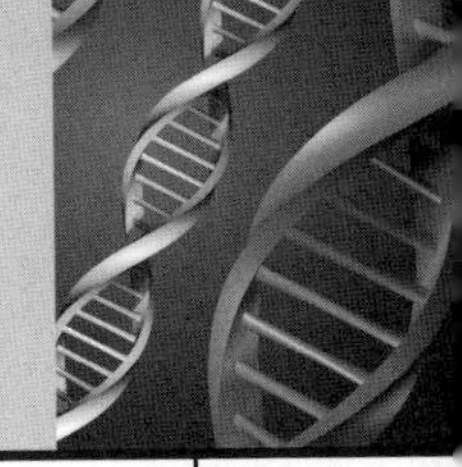

Q

When scientists first set out to map the human genome, they estimated that humans have about 100,000 genes, or more — maybe even 150,000! We're such complex creatures, it would make sense for us to have that many genes. The idea made sense at the time, but it proved to be wrong, wrong, wrong. Over the years, the estimated number of human genes has been trimmed about as often as some people cut their hair. Before it gets cut any farther, or grows again . . .

What has more genes than a human?

a) chicken

b) corn

c) fruit fly

d) mouse

The Map of Your Genes

What has more genes than a human?

a) chicken

b) corn

c) fruit fly

d) mouse

CORRECT ANSWER:

b) corn

According to recent estimates, humans have anywhere from 20,000 to 25,000 genes, roughly the same number as chickens. Now, just because we have about as many genes as chickens, doesn't mean we're all that closely related to them — unless you consider eating them to be a close relationship. In an evolutionary sense, we're much closer to mice who also have about as many genes as us. Fruit flies have roughly half as many as we do, but an ordinary ear of corn, or maize, has 45,000 genes — about twice as many as a human! Will you ever be able to look at a bowl of corn flakes in the same way again? Maybe it'll make you wonder about why corn needs so many genes and you don't.

The World's Most Successful Y Chromosome

Q

You probably know who your father is, and your grandfather too. But if you put about 200 greats in front of grandfather, would you be able to name that ancient ancestor? If you're a male, the secret of your forefather's identity is coiled up inside your Y chromosome, which is passed down in a direct line from father to son. One man's Y chromosome was so successful that he has millions of living descendents today. About one in every 200 males on the planet descended from his direct line. Are you one of them?

Who was this potent patriarch?

a) Attila the Hun

b) Chaka Khan

c) Genghis Khan

d) Kublai Khan

The World's Most Successful Y Chromosome

Who was this potent patriarch?

a) Attila the Hun

b) Chaka Khan

c) Genghis Khan

d) Kublai Khan

CORRECT ANSWER:

c) Genghis Khan

If you're male, and your father's line comes from anywhere in a huge chunk of Asia bounded by Manchuria in the east, and Uzbekistan and Afghanistan in the west, you could be descended from Genghis Khan. With more than 16 million male descendents in his direct line, he has 800,000 times as many male descendents as an average male who lived in his day — and that's not even counting the men and women who descended from his daughters. Genghis Khan conquered the world with his Mongol horde, and with his genes. The experts think that, one way or another, virtually everyone living near the Asian steppe probably has a bit of Genghis Khan in his or her genes.

Q

Some people can't watch a movie without a heaping helping of popcorn. Once the munching starts, it's hard to stop. Is it gluttony? Or are there cues we're unaware of that make us eat more? To find out if the size of the container influences how much people eat, food psychologist Brian Wansink gave away free popcorn. Moviegoers received one of a medium or large tub of fresh popcorn, or a medium or large tub of stale, five-day-old popcorn. Of the people who got fresh popcorn, those with the large tubs ate more than 50% as much as those with medium-sized tubs, but what about the ones who got the popcorn that was five days old?

How much stale popcorn did they eat from the large tub?

a) less than people eating from the medium tub

b) more than people eating from the medium tub

c) the same amount as people eating from the medium tub

d) they refused to eat any after they'd tasted it

You Can't Have Just One

How much stale popcorn did they eat from the large tub?

a) less than people eating from the medium tub
b) more than people eating from the medium tub
c) the same amount as people eating from the medium tub
d) they refused to eat any after they'd tasted it

CORRECT ANSWER:
b) more than people eating from the medium tub

Even when the popcorn was terrible, people with large tubs ate 45% more than those with medium tubs. There's something about bigger packages that compels us to keep eating. People consume at least 50% more eating from bigger containers than from smaller ones, and they're not even aware that they're doing it. It's not a problem if you're eating plain, air-popped popcorn, which only has about 30 calories a cup, but the calories add up with most other snack foods. Dr. Wansink recommends using smaller plates or bowls. They trick your brain into thinking there's more food, and if you think there's more food, you're more likely to feel satisfied with less.

No More Scraping the Bottom of the Barrel

Q

Maybe you're in the mood for some potato chips to enjoy with your flick. So, you grab a bag, rip it open, and dig in. Potato chips didn't always come in bags, you know. A hundred years ago, grocers sold them out of glass cases, or big, wooden barrels. People loved them, except for the broken, soggy, and stale ones scraped off the bottom of the barrel. In 1926, a clever chip maker came up with something that not only kept chips fresh longer, but also make them easier to sell and to eat — individual bags.

What were the bags made of?

a) cheap cloth stitched by hand
b) folded newspaper tied up with string
c) rice paper glued together with flour paste
d) sheets of waxed paper ironed together

No More Scraping the Bottom of the Barrel

What were the bags made of?

a) cheap cloth stitched by hand

b) folded newspaper tied up with string

c) rice paper glued together with flour paste

d) sheets of waxed paper ironed together

CORRECT ANSWER:

d) sheets of waxed paper ironed together

Laura Scudder, who worked in her family's chip business in California, came up with the idea of ironing together the edges of sheets of wax paper to form a bag. Female workers made the bags at home at night. The next day, at work, they filled them with chips, and sealed the tops with a warm iron. Goodbye to scraping the bottom of the barrel, and hello to self-serve snacks in disposable packaging.

What Your Favourite Comfort Food Says About You

Q

Maybe you've had a rough day and you crave comfort food. It makes you feel better, and makes movie time even more enjoyable. Our favourite foods tend to be holdovers from childhood. Comfort foods are usually things that we grew up eating and associate with pleasurable feelings. When we eat them, they bring back those feelings of comfort. Your favourite comfort food says something about you, and food psychologist Brian Wansink thinks he knows what it's saying. Who says that scientists don't have a sense of fun?

If your favourite comfort food is mac and cheese, what does that say about you? You're . . .

a) a straight shooter
b) flexible
c) nostalgic
d) sensitive

What Your Favourite Comfort Food Says About You

If your favourite comfort food is mac and cheese, what does that say about you? You're . . .

a) a straight shooter
b) flexible
c) nostalgic
d) sensitive

CORRECT ANSWER:

c) nostalgic

According to Professor Wansink, if mac and cheese gives you comfort, you're probably the nostalgic type. Your memories of the food bring you as much comfort as the food itself. You're low key, self reliant, and serene. Straight shooters find comfort in meatloaf and mashed potatoes, while flexible types turn to ice cream sundaes for an emotional boost. The sensitive types, meanwhile, hunker down with a chocolate brownie that evokes nurturing feelings. Spaghetti and meatball fans are family oriented, with traditional values. Are you unconventional? Do you like apple pie? Few people choose it as a comfort food these days, but those who do tend to be creative types. If your favourite food isn't listed, feel free to come up with your own description of what it says about you. There are no wrong answers.

Solar Pizza

Q

Maybe you're craving slice of pizza, dressed with your favourite toppings, to enhance your viewing pleasure. Here's something to think about while you're enjoying your special slice. What if the sun was a pizza, and the Earth was one of the toppings? Which topping would be as wide across as our planet?

If the sun were a 36-centimetre (14 in) pizza, the Earth would be a . . .

a) tomato slice
b) pepperoni slice
c) hot chili pepper seed
d) grain of salt

Solar Pizza

If the sun were a 36-centimetre (14 in) pizza, the Earth would be a . . .

a) tomato slice
b) pepperoni slice
c) hot chili pepper seed
d) grain of salt

CORRECT ANSWER:
c) hot chili pepper seed

If the sun is a pizza 36 centimetres (14 in) across, then our planet is a chili pepper seed in comparison. The sun's diameter is about 109 times that of the Earth. If you lined up 109 chili pepper seeds with a diameter of 3.3 millimetres (0.13 in), they'd form a row about 36 centimetres (14 in) long. If the tomato slice were the planet, the pizza would have to be about 5.5 metres (18 ft) across! The pepperoni slice is also too big. If it represented the earth, the sun pizza would be more than 3 metres (11 ft) in diameter. A regular grain of salt is way too small to be the Earth, but if we're talking about a grain of coarse salt, about 1 millimetre (0.04 in) across, it could stand in for the moon.

Q

Apollo 11 was about half a minute away from potential disaster during the first moon landing. On July 20, 1969, the lunar module, Eagle, came close to running out of fuel while the astronauts searched for a safe place to land. They touched down in the Sea of Tranquility in the nick of time. About six and a half hours later, Neil Armstrong and Buzz Aldrin were the first men to walk on the moon. In 1972, Apollo 17's Gene Cernan and Harrison "Jack" Schmitt were the last. Twenty-nine astronauts flew in Apollo spacecraft, and of those, twenty-four flew around the moon, but not all of them landed. Have you ever wondered how many sets of footprints there are on the moon?

How many men have left their footprints on the moon?

a) 7

b) 11

c) 12

d) 14

Footprints in the Moondust

How many men have left their footprints on the moon?

a) 7
b) 11
c) 12
d) 14

CORRECT ANSWER:
c) 12

A dozen men have left their footprints on the moon. Of that lucky 12, six also got to ride in lunar rovers. Their footprints and moon buggy tracks should still be there now, and could remain intact for millions of years to come, barring unforeseen cosmic or lunar events, or moon tourists. The men who left tracks on the moon were Neil Armstrong, Buzz Aldrin, Pete Conrad, Alan Bean, Alan Shepard, Edgar Mitchell, Dave Scott, Jim Irwin, John W. Young, Charlie Duke, Gene Cernan, and Harrison Schmitt, who was the only scientist of the 12. If all goes according to plan, humans should be back on the moon by 2020, about 50 years after the last astronauts left their footprints behind.

Moon Buggies

Q

Can you imagine being the first to drive a car where no man had driven before — on the surface of the moon? What a rush! It wasn't exactly a car though. It was a lunar rover, called, somewhat unimaginatively, Rover-1. Two happy astronauts were the first to drive a moon buggy to work. They could zip along at top speeds of 10 to 11 km/h (5.5 to 6 mph) on flat areas, but had to slow down to half that speed on turns and while driving over craters.

Who was the first astronaut to drive on the moon?

a) Buzz Aldrin
b) Neil Armstrong
c) Dave Scott
d) John Young

Moon Buggies

Who was the first astronaut to drive on the moon?

a) Buzz Aldrin
b) Neil Armstrong
c) Dave Scott
d) John Young

CORRECT ANSWER:

c) Dave Scott

Apollo 15 was the first mission to carry a lunar rover, and Dave Scott was the first man to drive it. Jim Irwin was next. For three days, they drove around near Mount Hadley, on the edge of the Apennine Mountains (named for Italy's mountain range with the same name). The first man to pop a wheelie on the moon was Apollo 16's John Young, who was the third man to drive on the moon. All three lunar rovers are still on the moon's surface, exactly where the astronauts parked them at the end of their last shift.

Moon Ball

On the third moon landing, Apollo 14's Alan Shepard and Ed Mitchell collected more moon rocks than previous Apollo missions, a total of 42 kilograms (94 lb). They were able to collect more because they were the first to use a handcart designed to carry their tools and transport the moon rocks. It worked for the most part, but when it couldn't negotiate a 15-degree slope on a big crater, the astronauts had to carry the cart to the rock collecting site. Despite the heavy workload, near the end of their last moonwalk, the astronauts found time for fun, and another first.

Q

What was the first ball on the moon?

a) football
b) baseball
c) basketball
d) golf ball

Moon Ball

What was the first ball on the moon?

a) football

b) baseball

c) basketball

d) golf ball

CORRECT ANSWER:

d) golf ball

Alan Shepard wanted to be the first man to drive a golf ball on the moon. To everyone's amusement, and in the interest of demonstrating the physics of low gravity, he attached a six iron to the bottom of a sampling instrument, and swung at a couple of balls. His best drive sent a golf ball into a distant crater. Shepard said it went for "miles and miles." For reasons known only to himself, Ed Mitchell grabbed a staff and threw it like a javelin. It landed in the same crater, and went a hair farther than the golf ball. Shepard brought his six iron back to Earth. It's on display at the U.S. Golf Association Hall of Fame in New Jersey. His two golf balls, as well as Mitchell's "javelin," are still on the moon.

The Real Man in the Moon

Footprints and tread marks aren't the only things humans left on the moon. Along with tons of hardware and technology, including probes, rocket stages, rovers, tools, equipment, and experiments, there are personal items too. One of them is about as personal as it can get: someone's "cremains." An ounce of human ashes hitched a ride on Lunar Prospector, which orbited the moon for 19 months searching for water. For its final experiment, and spectacular grand finale, the probe targeted a shadowed crater near the south pole, and hurled itself and the cremains into it.

Who is "buried" on the moon?

a) Gene Autry, singing cowboy
b) Gene Roddenberry, *Star Trek* creator
c) Gene Shoemaker, geologist
d) Virgil "Gus" Grissom, astronaut

The Real Man in the Moon

Who is "buried" on the moon?

a) Gene Autry, singing cowboy
b) Gene Roddenberry, *Star Trek* creator
c) Gene Shoemaker, geologist
d) Virgil "Gus" Grissom, astronaut

CORRECT ANSWER:

c) Gene Shoemaker, geologist

Geologist Gene Shoemaker had dreamt of going to the moon, but couldn't be an astronaut for medical reasons. So, instead, he trained astronauts, and organized the geological activities planned for the moon missions. He was a crater specialist, a pioneer of astrogeology, and discovered many comets. He headed up the 1994 Clementine mission, which searched for water in shadowed craters around the south pole — the same area where he and his ride, Lunar Prospector, slammed into the moon's surface in 1999. Yes, there really is a man in the moon, and his name is Gene Shoemaker.

Q

Half the cars in America were Model Ts in 1918. In the early days of the automobile trade, no one came out with a new model every year. They sold the same model for years. A decade after the Model T was introduced, there were more of them than any other car, and many, if not most, were black. It was the only colour available from 1915 to 1925. Why? Good question.

Why were Model Ts only made in black for 11 years?

a) as a gesture of mourning for Ford's deceased son

b) black paint dried faster

c) Ford was colour-blind

d) it was the most popular colour

Paint it Black

Why were Model Ts only made in black for 11 years?

a) as a gesture of mourning for Ford's deceased son
b) black paint dried faster
c) Ford was colour-blind
d) it was the most popular colour

CORRECT ANSWER:

b) black paint dried faster

Business was booming, and black paint dried faster than other colours. Ford had the most efficient assembly line in the world, and turned out a new car every 98 minutes, but it was still a struggle to keep up with demand. The faster the paint dried, the more cars the plant could churn out. The Model T was "the car" for decades, but its run finally ended in 1927 when the last Model T, number 15,007,034, left the building. More Model Ts were made than any other car in the world, and that record stood for 45 years, right up until the 15,007,035th Volkswagen Beetle rolled off its assembly line.

The Top 10 Bestselling Cars of All Time that Never Underwent a Major Redesign

1. Volkswagen Beetle
2. Ford Model T
3. Lada Riva
4. Fiat Uno
5. Renault 4
6. Mini
7. Peugot 206
8. Peugot 205
9. Ford Model A
10. Hindustan Ambassador

The Mercedes Era

Q

The Mercedes was born thanks to the demands of Emil Jellinek. He wanted a revolutionary sports car, "not a car for today or tomorrow, but for the day after tomorrow." His demands included a longer wheelbase and wide track for stability, a lower centre of gravity, and electric ignition. The Mercedes that Daimler and Maybach built for him amazed car fans all over the world in 1901. Jellinek reached the then-incredible speed of 60 km/h (37 mph) to win the Nice races with ease, beating cars in all capacity classes. The Director of the French Automobile Club announced that the world had entered the Mercedes era. You probably know about the car, but what about other things named Mercedes?

What was NOT named Mercedes?

a) Emil Jellinek

b) Emil Jellinek's daughter

c) Emil Jellinek's mother

d) Emil Jellinek's properties

What was NOT named Mercedes?

a) Emil Jellinek

b) Emil Jellinek's daughter

c) Emil Jellinek's mother

d) Emil Jellinek's properties

CORRECT ANSWER:

c) Emil Jellinek's mother

Emil Jellinek's mother was named Rosalie. Jellinek's love for the name began after his first daughter was born, and named Mercedes. He thought it was an exotic, attractive, and lucky name, and he stipulated that Daimler and Maybach were to name their new car Mercedes. He named his race team and all the properties he owned Mercedes, including three separate Villa Mercedes. At one point he decided that the Mercedes engine's future was on the water, so he put one on each of his yachts, and called every single yacht — you guessed it — Mercedes. But it gets even better! In 1903, when he was 50, he legally changed his name to Emil Jellinek-Mercedes. He joked that it was the first time a man had taken his daughter's name. No kidding.

Car Kings

Q

Two of the pioneers of the car industry, Karl Benz and Gottlieb Daimler, were born about 60 kilometres (37 miles) apart in southern Germany. Karl Benz was born and educated in Karlsruhe, where Daimler and his partner Wilhelm Maybach worked for a manufacturer for two years. To support their families, and to finance the development of their engines, these future car kings had other jobs in the early days, and worked on their engines in their spare time. By 1886, Benz was in Mannheim testing a motorized tricycle he'd built, while Daimler and Maybach were in Stuttgart test driving a horse carriage rigged with one of their engines. Fast-forward 25 years to the merger of Daimler and Benz's companies. The merger was a good fit, but what about the car kings themselves?

What was Daimler and Benz's relationship? They . . .

a) were rivals in school

b) worked together in Karlsruhe and became friends

c) were related by marriage

d) had no relationship

What was Daimler and Benz's relationship? They . . .

a) were rivals in school

b) worked together in Karlsruhe and became friends

c) were related by marriage

d) had no relationship

CORRECT ANSWER:

d) had no relationship

Daimler's and Benz's companies merged during rough financial times in Germany in 1926. Daimler had already been dead for 26 years, and Benz was retired. Despite the fact that their birthplaces were close together, and they lived and worked in the same neck of the woods, there's no evidence that the two ever met. Based on how different their engine designs were, it's possible that they weren't even interested in one another's work. Even though the founders of the two companies weren't involved in the merger, or maybe because they weren't involved, the new company, Daimler-Benz, proved to be incredibly successful. Their most popular luxury vehicle, the Mercedes-Benz, is still an automotive object of desire.

Patented Success Stories

The men who rose to be the kings of the motor trade got there, not just by having good ideas, but also by getting patent protection for their inventions. More than a century ago, thousands of inventors all over the world were furiously working away on different kinds of engines and cars. Sometimes more than one inventor came up with the same idea. The ones who patented their inventions first were the ones who got the bragging rights, and if they had good lawyers, money for the use of their patented ideas. They were a creative bunch and came up with all kinds of innovations.

Match the car kings with their achievements.

a) Benz
b) Daimler and Maybach
c) Diesel
d) Ford

1. A charcoal briquette, an ethanol-fueled plastic-bodied car, set land speed record
2. First commercial engine at a brewery, solar-powered air engine, compression ignition engine
3. First motorcycle, first motorboat, the Phoenix engine
4. First commercially available automobile, the truck, race car engine

Patented Success Stories

Match the car kings with their achievements.

a) Benz

b) Daimler and Maybach

c) Diesel

d) Ford

1. A charcoal briquette, an ethanol-fueled plastic-bodied car, set land speed record
2. First commercial engine at a brewery, solar-powered air engine, compression ignition engine
3. First motorcycle, first motorboat, the Phoenix engine
4. First commercially available automobile, the truck, race car engine

CORRECT ANSWERS: a-4, b-3, c-2, d-1

Benz was the first to advertise and sell his car, the three-wheeled Benz Patent Motorwagen, in 1888. He also designed the first truck, and patented an engine design still used in some high performance race cars. Daimler and Maybach were the first to attach their engine to a bicycle, and a boat. Their Phoenix engine powered the car that won the world's first car race in 1894. Rudolf Diesel invented the compression ignition engine, a.k.a. the Diesel engine, and his first commercial engine powered a brewery in the U.S. He also patented a solar-powered air engine. Ford and his brother-in-law E.G. Kingsford created a charcoal briquette using sawdust and wood chips from the car factory. Ford held 161 patents, including one for an ethanol-fueled car with a body made of soybean plastic. In 1904, he set a world land speed record of 147 km/h (91.3 mph) but it was broken a couple of months later.

Q

Inside every chicken beats the heart of a wild Red Jungle Fowl. Chickens are domesticated jungle fowl, but that hasn't changed their wild instincts. In the jungles of Asia, wild chickens live in flocks, and spend most of their time hunting for food and mating. A closer look at the sex lives of jungle fowl reveals a strategy that seems superficial: they choose their mates based on looks. Studly roosters go for the hottest chicks and vice versa. Most of us have some idea of what's hot for humans, but what could possibly be hot, or not, for chickens?

What makes chickens attractive to the opposite sex?

a) a big beak

b) a voluptuous comb

c) small ears

d) the yellowest feet

Chicken Beautiful

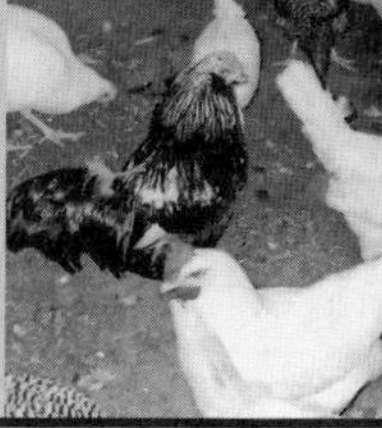

What makes chickens attractive to the opposite sex?

a) a big beak

b) a voluptuous comb

c) small ears

d) the yellowest feet

CORRECT ANSWER:

b) a voluptuous comb

The comb is the fleshy red tissue on top of a chicken's head, and it's what makes a chicken hot, or not. The bigger, brighter, and more voluptuous the comb, the more attractive and desirable the chicken. That goes for males and females, both wild and domesticated. Now, before you write the chicken off as vain and shallow, consider this: only healthy chickens have healthy-looking combs, and a robust mate is a chicken's best shot at producing healthy offspring. Chickens may not be rocket scientists, but when it comes to the survival of their species, they've got great instincts.

Gamy Roosters

Q

Roosters live for two things: fighting and fertilizing. A rooster's got to do what a rooster's got to do to father as many offspring as he can, and preferably more than the other roosters. Some part of his tiny brain weighs the odds of mating success in different situations, and prompts the rooster's behaviour. Sometimes it's in his best interest to give a lot of sperm to a hen, and sometimes it's not. It all depends on the circumstances, and his status in the flock.

When will a dominant rooster give a hen the most sperm? When . . .

a) she's one of many hens

b) she's the only hen

c) there are no other roosters present

d) there are other roosters present

Gamy Roosters

When will a dominant rooster give a hen the most sperm? When . . .

a) she's one of many hens

b) she's the only hen

c) there are no other roosters present

d) there are other roosters present

CORRECT ANSWER:

d) there are other roosters present

The amount of sperm a dominant wild rooster contributes depends on whether there are other males around, and how many of them there are. When rivals are present, a dominant rooster gives more sperm, in a bid to win the fertilization competition. Low-ranking roosters do the opposite, reducing their sperm output when rivals rule the roost. Maybe they're waiting for a better, less competitive situation to come up. If there are no rivals around, roosters give the minimum amount of sperm required for fertilization, but even in that situation, they give more to the attractive hens than to the homely ones.

Egg Layers

Chickens are egg-laying machines. They lay more eggs than any other bird. With all that mating going on, most eggs get fertilized in the wild. Laying them fulfills the hen's biological imperative to bring more chicks into the world, and domestic chickens can't turn that egg-laying instinct off, even when there are no roosters around to fertilize their eggs. As long as the eggs are removed, hens just keep on laying them. In the right circumstances, a hen will lay up to 300 eggs a year. Now you know why hens lay so many eggs, and you know how many, but do you know how those eggs form? Which part of the egg comes first: the shell, the white, or the yolk? That's the question . . .

In which order is an egg formed inside the hen?

a) shell, white, yolk
b) shell, yolk, white
c) white, yolk, shell
d) yolk, white, shell

Egg Layers

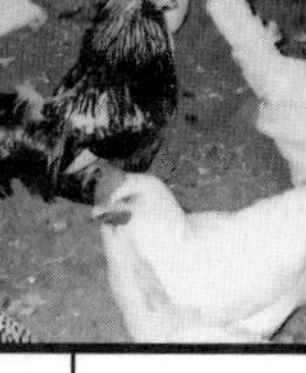

In which order is an egg formed inside the hen?

a) shell, white, yolk
b) shell, yolk, white
c) white, yolk, shell
d) yolk, white, shell

CORRECT ANSWER:

d) yolk, white, shell

When a hen ovulates, the yolk is released and starts its journey through the more than 60-centimetre (2 ft) long oviduct. First the yolk gets covered with a thin membrane and structural fibres, and then in layers of egg white, or albumin. As it moves through the long, spiralling oviduct, the fibres twist and form into two thick ropes that keep the yolk centered in the egg. The final step takes place in the shell gland, where the shell forms. This stage takes the longest, about 20 hours. If it's a brown egg, there's an extra step during the last few hours of shell formation when the colour is added. The brown pigment is a breakdown product of the hemoglobin in the chicken's blood. Other than pigment on the shell, there's no difference between brown and white eggs.

Headless Chicken

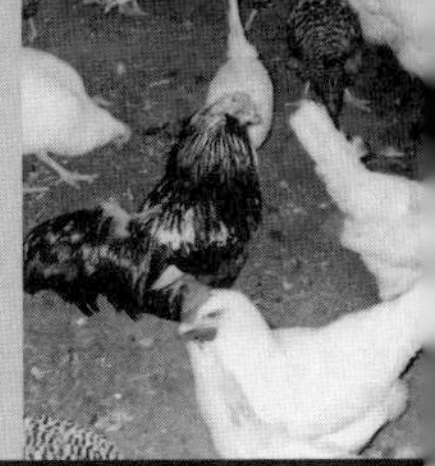

Q

Have you ever heard the saying "running around like a chicken with its head cut off"? Maybe you've seen cartoons with some hapless animated chicken running around frantically looking for its disembodied head. Although it sounds like cartoon fodder, or an urban myth, it's not. Headless chickens can survive in certain circumstances. How is that possible? And even if it is possible, how long could a chicken possibly live without a head? You just had to ask, didn't you?

How long has a chicken survived after its head was cut off?

a) 18 hours
b) 18 days
c) 18 months
d) 18 years

Headless Chicken

How long has a chicken survived after its head was cut off?

a) 18 hours
b) 18 days
c) 18 months
d) 18 years

CORRECT ANSWER:
c) 18 months

A chicken who was supposed to be dinner became a headless celebrity instead. Mike the Rooster's journey began with a botched decapitation. The thinking parts of his brain were chopped off, but an ear and most of his brain stem remained intact. The brain stem controls basic physical functions, so Mike could do most things that didn't require having a head. He became an instant celebrity when *Life* magazine covered him in 1945. Soon Mike had a manager and was touring America. During his 15 minutes of freak show fame, Mike made a small fortune — more than $50,000 a month in today's dollars. Then, when he was 18 months old, tragedy struck, and neither his fame nor his fortune could save him. After a gig, back in the motel room he shared with his owners, he choked on his own mucus in the middle of the night. Whether his owners ate him at that point is unknown.

Q

Women have a very special relationship with chocolate. Of the females in an American survey, 93% said that they ate it. More than half said that chocolate made them happy. In a British survey, 66% of the women described chocolate as a mood enhancer, but only 38% of the men knew that chocolate seduces the brain into releasing feel-good chemicals called endorphins. They act like drugs to kill pain, reduce stress, and induce feelings of euphoria. Is it any wonder women have a love affair with chocolate?

How many women said they'd rather have chocolate than sex?

a) 71%
b) 52%
c) 25%
d) 17%

Will that Be Sex or Chocolate?

A

How many women said they'd rather have chocolate than sex?

a) 71%
b) 52%
c) 25%
d) 17%

CORRECT ANSWER:
b) 52%

About 52% of the women in the British survey said that they'd rather indulge in chocolate. Imagine the disappointment of the men surveyed: 87% of them said they'd rather have sex. So, what's the big deal with chocolate? One woman may have hit the nail on the head when she said that chocolate provides "guaranteed pleasure and never disappoints." Is there anything else on earth you can say that about? Half the men surveyed said that they regularly gave chocolate as birthday and thank-you gifts. Most women on the receiving end of chocolate probably appreciate it, but what (if anything) happens next depends on the man's relationship with the woman, and the woman's relationship with chocolate.

In Love with Chocolate

Have you ever fallen in love? You feel stimulated and excited, maybe even giddy, but with a sense of well-being. If you like that feeling, you can try to fall in love on a regular basis, or you can try eating chocolate. Falling in love steeps your brain in a stimulating chemical called phenylethylamine, a.k.a. the "love drug," which just happens to be found both in your brain and in chocolate. When you eat chocolate, the "love drug" makes its way to your brain and works its magic. Phenylethylamine revs up your brain, but it's not the only stimulant in chocolate. You've probably heard there's caffeine in it too, but do you know how much? Let's say that a cola is a 355-millilitre (12 oz) can, and that the other beverages are 237-millilitre (8 oz) cups.

How much caffeine is in 28.5 grams (1 oz) of chocolate? About as much as . . .

a) brewed coffee

b) cola

c) decaffeinated coffee

d) tea

In Love with Chocolate

How much caffeine is in 28.5 grams (1 oz) of chocolate? About as much as . . .

a) brewed coffee
b) cola
c) decaffeinated coffee
d) tea

CORRECT ANSWER:
c) decaffeinated coffee

There isn't much caffeine in chocolate. Some experts think that there isn't any caffeine, but there definitely is another stimulant, theobromine, that's similar to caffeine in some ways. Apparently, if you eat enough chocolate, the theobromine can keep you awake at night. Well, I confess to having consumed mass quantities of chocolate, and it's never kept me awake, but your mileage may vary. Theobromine is a kinder, gentler stimulant than caffeine. It doesn't deliver a jolt, and lasts longer. The word theobromine comes from the cacao plant's botanical name, *Theobroma*, which means food of the gods. It was named by the famous Swedish naturalist Linnaeus. Food of the gods, eh? Do you think he was in love with chocolate?

High on Chocolate

Q

Chocolate is a drug. Its active ingredients alter the chemistry of the brain and body. Everything you eat or drink alters your brain and body and is a drug in that sense, but chocolate has some extra special mind-altering qualities. Scientists know of nearly 400 chemicals in chocolate, but there are still surprises hidden deep inside the dark mass of edible delight. Three compounds that resemble a brain chemical called anandamide were discovered not long ago. Anandamide induces a marijuana-like high because it fits into the same brain receptors as THC, the active ingredient in cannabis. If the anandamide in chocolate mimics THC in the brain, is it possible to get high on chocolate?

How much chocolate would a 59-kilogram (130 lb) person have to eat in one sitting to get high? Approximately . . .

a) 0.9 kg (2 lb)
b) 4.5 kg (10 lb)
c) 9 kg (20 lb)
d) 11.4 kg (25 lb)

High on Chocolate

How much chocolate would a 59-kilogram (130 lb) person have to eat in one sitting to get high? Approximately . . .

a) 0.9 kg (2 lb)
b) 4.5 kg (10 lb)
c) 9 kg (20 lb)
d) 11.4 kg (25 lb)

CORRECT ANSWER:
d) 11.4 kg (25 lb)

Even though 11.4 kilograms (25 lb) of chocolate contain enough chemicals to induce a high, you can't really get high on chocolate because you can't eat enough. Neuroscientists studied the effects of too much chocolate by imaging the brains of chocoholics as squares of chocolate melted slowly in their mouths. The subjects consumed anywhere from 40 to 170 grams (1 to 6 oz) of chocolate during the experiment. At first, the pleasure centres of their brains lit up in the areas activated by addictive drugs, such as cocaine, but when they ate too much, the thrill was gone. The more they had to eat, the less pleasurable it became until, eventually, forcing themselves to eat the chocolate felt like punishment. Now imagine trying to make yourself eat 100 times the amount that made the chocoholics suffer. I bet your brain wouldn't like it.

Chocoholics Live Longer

Chocolate lovers live longer. That's the conclusion of two different studies: one of 8,000 Harvard University graduates, and the other of a group of elderly Dutch men. In the Dutch study, chocolate eaters had lower blood pressure, and healthier cardiovascular systems than those who didn't eat it. The men who ate the most chocolate on a daily basis lived the longest. The researchers think that the chocolate advantage comes from antioxidants called polyphenols. It's thought that they contribute to good health, and maybe even longevity, by protecting the body against cell damage. Cacao beans are rich in polyphenols. How rich?

Forty grams (1.4 oz) of dark chocolate packs about the same polyphenol punch as . . .

a) 1 glass of red wine

b) 2 servings of berries

c) 5 servings of fruits and vegetables

d) 10 servings of fruits and vegetables

Chocoholics Live Longer

Forty grams (1.4 oz) of dark chocolate packs about the same polyphenol punch as . . .

a) 1 glass of red wine

b) 2 servings of berries

c) 5 servings of fruits and vegetables

d) 10 servings of fruits and vegetables

CORRECT ANSWER:

d) 10 servings of fruits and vegetables

Cacao beans, the main ingredient in chocolate, rule when it comes to packing the most polyphenols. A 40-gram (1.4 oz) piece of milk chocolate contains as much as a glass of red wine, or five servings of fruit and veg. The same amount of dark chocolate has twice the cacao, and twice the polyphenols — the amount you'd find in 10 servings of fruit and veg, or two glasses of red wine. But before you turn chocolate into a food group, consider that chocolate contributes mostly calories and fat to your diet. If you wanted to burn off the calories from 40 grams (1.4 oz) of chocolate, you'd have to bicycle, or walk, for about half an hour. The right dose of chocolate might cheer you up, and the polyphenols might even contribute to your health somewhat, but it's still not what you'd call a health food.

Q

You have to go — badly — and the only option is a public restroom. The horror! You start to wonder how many people have used the facilities before you, and what smelly things they might have left behind. Maybe you're worried about the vicious viruses and bacteria lurking in and around the toilets, ready to glom onto your unsuspecting flesh and infect you with something horrible. But, fortunately, you can choose from among four public restrooms, all within the same short walking distance.

Where is the cleanest public restroom?

a) airport

b) bus station

c) fast food restaurant

d) hospital emergency room

The Horror

Where is the cleanest public restroom?

a) airport

b) bus station

c) fast food restaurant

d) hospital emergency room

CORRECT ANSWER:

d) hospital emergency room

According to The Germinator, a.k.a. Dr. Chuck Gerba, who's tested more toilets than you'll ever want to imagine, the cleanest restrooms are in hospital emergency rooms, followed by fast food restaurants. That's good news, considering that you'll almost always be within walking distance of a fast food place, if not an ER. Airport, bus station, and gas station restrooms are the worst. Testing them is a dirty job that sometimes attracts the kind of attention The Germinator could do without. Once, when he was on his knees in front of a public toilet, he heard a tapping on the stall. It was a policeman. "Are you the only one in there?" he asked. The Germinator answered, "I'm a scientist. I'm doing research." "Yeah, right," the cop retorted, "I arrested one of you last week."

The Cleanest Stall of All . . . Shhh!

Q

So now you're standing in the public restroom trying to decide which stall to choose. Unfortunately, The Germinator isn't there testing them, so he can't tell you which one to pick. You can do a visual inspection to find the cleanest one, but that means touching doors unnecessarily, not to mention it could be scary, tedious, and time-consuming, and you really, really have to go. Did you know that there's a shortcut for locating the cleanest toilet? It's not a secret, but it might as well be, because most people don't know about it — and I think we should keep it that way.

Which stall is likely to be the cleanest?

a) the first stall on the left

b) the middle stalls

c) the second stall on the right

d) the stall with lots of toilet paper on the floor

The Cleanest Stall of All . . . Shhh!

Which stall is likely to be the cleanest?

a) the first stall on the left

b) the middle stalls

c) the second stall on the right

d) the stall with lots of toilet paper on the floor

CORRECT ANSWER:

a) the first stall on the left

The more stalls there are the better, but some still get used more than others. To find out which ones were used the most, and the least, Dr. Gerba numbered the squares of toilet paper in all the stalls in a public restroom. He discovered that the ones in the middle had the highest volume of traffic, and the first stall had the lowest. If the toilets are laid out in a way that lets you choose between the first toilet on the left, or the first on the right, take the first on the left. More people go right. But don't tell anyone. The more people who know, the less chance it'll still be the cleanest when you or I need to use it next.

To Wash or Not to Wash

Q

Mission accomplished, you step out of the cubicle. Now you're in front of the sinks wondering whether you'd be better off washing your hands or not. Would washing lead to more germs on your hands or less?

What should you do?

a) Don't wash your hands! Just get out as fast as you can.

b) Rinse hands with cold water, then dry them on your hair.

c) Scrub with soap, rinse well, use a paper towel to turn off the taps and open the door.

d) Use hand sanitizer.

To Wash or Not to Wash

What should you do?

a) Don't wash your hands! Just get out as fast as you can.

b) Rinse hands with cold water, then dry them on your hair.

c) Scrub with soap, rinse well, use a paper towel to turn off the taps and open the door.

d) Use hand sanitizer.

CORRECT ANSWERS:

c) Scrub with soap, rinse well, use a paper towel to turn off the taps and open the door, and d) Use hand sanitizer.

Up until I spoke with The Germinator the only correct answer would have been choice c). That's how microbiologists wash their hands. It's a tried and true method that avoids some of the worst hot spots, or at least we thought it did. But recently, Dr. Gerba was more than a bit surprised to find poopy germs in the liquid soap in public restrooms. Wait a minute . . . isn't soap supposed to kill bacteria, or at least get them *off* your hands? How bacteria can thrive in soap dispensers is a mystery. Until the mystery is solved and a solution found, The Germinator has switched to hand sanitizer. If it's good enough for The Germinator, it's good enough to be a correct answer.

The 10 Most Contaminated Things in Public Restrooms

Q

Do you have a ritual for avoiding germs in public restrooms? Maybe you put toilet paper all over the seat first. Or perhaps you suspend yourself above the toilet, so that no part of you touches it. Maybe you flush it with your foot. You can't see the germs, but you know for sure that they're everywhere. Your first line of defence against germs is knowing where the worst ones live, and avoiding them. Some areas in public restrooms have lots of nasty bacteria, while other areas have relatively few.

Where would you find the most nasty bacteria in public restrooms?

a) floor of the stall
b) restroom door handle
c) tap handles
d) toilet seat

The 10 Most Contaminated Things in Public Restrooms

Where would you find the most nasty bacteria in public restrooms?

a) floor of the stall
b) restroom door handle
c) tap handles
d) toilet seat

CORRECT ANSWER:

c) tap handles

Dr. Gerba found the nastiest of the nasty bacteria only on the tap handles and sinks. Salmonella and shigella are pathogens, meaning they can make you really sick. They can be quite dangerous for the very young, the very old, and those who aren't healthy. Some people refuse to touch the exit door handle, but that's one area where you'll find the fewest germs. Another relatively germ-free zone is the top of the toilet seat in women's restrooms. The Germinator thinks it might be because more women than men wipe the seat before sitting down.

The Top 10 Most Contaminated Things in Public Restrooms

1. tap handles
2. sink
3. underside of the toilet seat
4. floor around the toilet
5. around the sanitary napkin disposal
6. inside the urinal
7. floor in front of the urinal
8. diaper changing table
9. toilet handle
10. liquid soap dispenser

Q

I don't have to tell you who Albert Einstein was, but you might not be as familiar with his one-time inventing partner, Leó Szilárd. He was one of Einstein's former students, and a brilliant physicist himself. Just like a starving artist obsessed with art, Szilárd was a starving physicist obsessed with science, and he needed money to support himself and his research. Einstein came to his rescue and joined him on a project. The two geniuses set out to invent something that they hoped would generate income for Szilárd to finance his future work.

What did Albert Einstein and Leó Szilárd invent together?

a) a microwave oven

b) a refrigerator with no moving parts

c) a solar-powered water heater/boiler system

d) an atomic clock

\$=es² (\$=EinsteinSzilárd²)

What did Albert Einstein and Leó Szilárd invent together?

a) a microwave oven
b) a refrigerator with no moving parts
c) a solar-powered water heater/boiler system
d) an atomic clock

CORRECT ANSWER:

b) a refrigerator with no moving parts

Einstein and Szilárd invented three different cooling technologies unlike anything we have today. None of them had moving parts, and all of them were energy efficient and very unusual. One was powered by the heat of a gas pilot light, another had an electromagnetic pump, and the third relied on evaporation and the pressure of tap water to keep things cool. During their seven-year collaboration, Einstein, a former patent clerk, registered dozens of patents, some of which were licenced by big companies. Even though none of their refrigerators were made commercially, Szilárd made enough money from the licencing royalties to finance his most significant research — the nuclear chain reaction that led to the world's first atom bomb.

Bombing the Earth's Magnetic Field

What would you say if I told you that three 1.7-megaton nuclear bombs were about to be exploded above the atmosphere, in our planet's magnetic field? Are you outraged? Relax. It's already happened. It wasn't a random act of kookiness either, but a top-secret American military project called Argus. The goal of the 1958 experiment was to create artificial radiation belts in an area of the Earth's magnetic field called the Van Allen Belts, which the military could then use to disrupt enemy communications satellites, or incoming missiles.

Who headed up Project Argus?

a) Air Force mechanic

b) amateur astronomer

c) elevator engineer

d) Lieutenant General's nephew

Bombing the Earth's Magnetic Field

Who headed up Project Argus?

a) Air Force mechanic

b) amateur astronomer

c) elevator engineer

d) Lieutenant General's nephew

CORRECT ANSWER:

c) elevator engineer

Nicholas Christofilos worked as an elevator engineer, but physics was his passion, and he taught himself everything there was to know about the motion of ions and electrons in magnetic fields. Project Argus was his idea, and the military put him in charge of it. The artificial radiation belts created by the blasts broke up after a few weeks. So, in 1962, Starfish Prime, a hydrogen bomb 1,000 times as powerful as the Argus bombs, was exploded in the Van Allen Belts. That blast created a radiation belt that lasted for years, and fried or damaged a handful of satellites, including the world's first commercial communications satellite, Telstar. More than a dozen nuclear bombs were detonated at high altitudes by the U.S. and the USSR, and the Van Allen Belts haven't been the same since. No one knows when, or if, they'll ever return to their former state.

Cooking with Genius

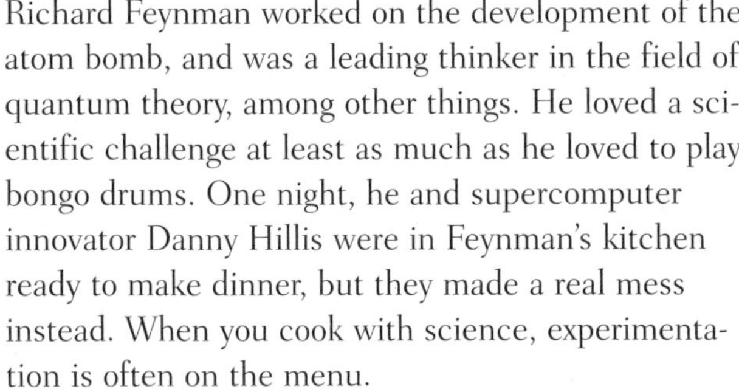

Richard Feynman worked on the development of the atom bomb, and was a leading thinker in the field of quantum theory, among other things. He loved a scientific challenge at least as much as he loved to play bongo drums. One night, he and supercomputer innovator Danny Hillis were in Feynman's kitchen ready to make dinner, but they made a real mess instead. When you cook with science, experimentation is often on the menu.

What were they trying to discover?

a) Does the buttered side of toast hit the floor more often?
b) What is the velocity of honey?
c) Why does bottled soda explode when mints are dropped into it?
d) Why does spaghetti shatter when you try to break it in half?

What were they trying to discover?

a) Does the buttered side of toast hit the floor more often?

b) What is the velocity of honey?

c) Why does bottled soda explode when mints are dropped into it?

d) Why does spaghetti shatter when you try to break it in half?

CORRECT ANSWER:

d) Why does spaghetti shatter when you try to break it in half?

If you hold a strand of spaghetti by the ends, and try to break it in half by bending it until it breaks, it shatters, often into three or more pieces. Danny Hillis asked Richard Feynman why that happened, and the experiment was on. Instead of cooking their pasta, they played with it and tried several different techniques for breaking it. After a few hours, they had broken spaghetti all over the kitchen, but had no dinner or good theory to explain why spaghetti shatters. Too bad they didn't have a high speed camera that could shoot 1,000 frames a second. That's what French scientists trying to solve the mystery used 20 years later. They think that the first break in the strand creates waves that travel through the pasta, which cause it to break up into three or more pieces. Despite having a high speed camera, they made a mess too.

A Psychologist and his Box

Q

B.F. Skinner studied behaviour, and is famous for ingenious animal experiments using his "Skinner Box," a cage rigged with a lever or button that released food when it was pressed. His experiments showed that animals could be trained to do all kinds of complicated tasks by using a combination of positive and negative reinforcements. Late in his career he trained pigeons to act as if they were self-aware, and could communicate with each other. He probably could have herded cats, but as far as we know, he didn't.

What else did B.F. Skinner NOT do?

a) sleep in a bright yellow plastic tank in his office

b) teach a pigeon to bowl

c) teach a rat to spend money

d) teach rats to guide warheads

A Psychologist and his Box

What else did B.F. Skinner NOT do?

a) sleep in a bright yellow plastic tank in his office

b) teach a pigeon to bowl

c) teach a rat to spend money

d) teach rats to guide warheads

CORRECT ANSWER:

d) teach rats to guide warheads

Dr. Skinner trained his lab rats to complete all sorts of complicated tasks, but he worked with 24 *pigeons* when America needed a missile guidance system during World War II. The birds were conditioned to keep a target in the centre of a screen by pecking at four levers (up, down, left, right). He envisioned three pigeons, each in a separate compartment of the missile's nose cone, their combined pecks steering the warhead to the target. Project Pigeon was tested, but in the end it didn't fly. The missile engineers and physicists wanted an electronic guidance system because it was the way of the future, but their early systems rarely hit a target. Maybe they should have gone for pigeon power. With today's technology, the pigeons wouldn't even have to leave their roosts.

Q

Our obsession with fast-moving things compels us to race whatever we can, wherever, and whenever we can. It thrills us to see the fastest of the fast compete. Animals in the wild don't race for sport, and they don't keep speed records. We more than make up for their lack of interest by timing anything that moves, or *really* moves — like the fastest living thing on the planet. Can you name the world speed record holder? I've heard different answers to that question. Maybe you have too. Let's say they're all clocked going at top speed.

Which one is moving the fastest?

a) Bdellovibrio (bacterium)

b) cheetah

c) mako shark

d) peregrine falcon

Fastest Living Thing on the Planet

Which one is moving the fastest?

a) Bdellovibrio (bacterium)

b) cheetah

c) mako shark

d) peregrine falcon

CORRECT ANSWER:

d) peregrine falcon

Peregrines are the fastest living things on the planet. They can plummet through the air at more than 300 km/h (186 mph) when stooping, or diving, for prey. Cheetahs, the next fastest, can run short distances at about 110 km/h (68 mph). The mako is the fastest shark, with a top speed of about 96 km/h (60 mph). The bacterium, travelling at more than half a metre (20 inches) per hour, is dead last when it comes to actual speed. But if you compare speed relative to body length, the bacterium is faster than even the peregrine. In attack mode, Bdellovibrio* can move at 100 times its body length per second. If we could move like that, we'd be roaring around as fast as a passenger jet.

* Bdellovibrio is pronounced dellovibrio, with a silent B.

Heaviest Baby Panda

The heaviest panda cub was born at the Wolong Giant Panda Research Center in China in 2006. He was Zhang Ka's first baby, and she set a record too, for the longest labour by a panda, 34 hours. The staff were minutes away from starting an operation to deliver the baby, but at the last minute, the wailing of the newborn baby panda stopped and delighted them. Zhang Ka, an average sized female panda, weighed 103 kilograms (227 lb), but her baby was a real heavy-weight.

How much did the panda cub weigh?

About as much as . . .

a) a newborn polar bear cub
b) a newborn teacup poodle
c) an official-size soccer ball
d) two sticks of butter

Heaviest Baby Panda

How much did the panda cub weigh?

About as much as . . .

a) a newborn polar bear cub

b) a newborn teacup poodle

c) an official-size soccer ball

d) two sticks of butter

CORRECT ANSWER:

d) two sticks of butter

The chubby baby panda weighed 218 grams (7.7 oz), almost as much as two sticks of butter. At birth, the average panda cub weighs about as much as one stick of butter, as does an average newborn teacup poodle. As an adult, the teacup poodle will weigh about 20 times as much as when it was born. The average baby panda increases its birth weight by about 1,000 times, and grows into a 100–150 kilogram (220–330 lb) adult. Will the chunky baby panda, named Bao Le, be a super-sized adult? We'll have to wait and see, and while we're waiting, consider this: if humans were less like dogs, and more like pandas, a three-kilogram (6.6 lb) baby would grow into an adult weighing three or four tonnes, about as much as an elephant.

The Hottest Stars

Q

The next time you're stargazing, how cool would it be if you could just look up and casually point out the hottest stars in the night sky? You don't need a degree in astronomy, a star map, or even a telescope. The secret of identifying which stars are hot, and which are not, is simple — colour. The hottest stars in the night sky twinkle blue or bluish white. Now that you know what to look for, do you know how hot the hottest of the hot are?

How hot is the surface of the hottest stars? About the same temperature as . . .

a) a bolt of lightning
b) inside a car's catalytic converter
c) inside a particle accelerator
d) the core of the stars

The Hottest Stars

How hot is the surface of the hottest stars? About the same temperature as . . .

a) a bolt of lightning
b) inside a car's catalytic converter
c) inside a particle accelerator
d) the core of the stars

CORRECT ANSWER:
a) a bolt of lightning

Blue stars may be among the hottest, but at the surface, their temperature is only about as hot as a bolt of lightning here on Earth, around 30,000°C (54,000°F). That's more than five times hotter than the surface of our sun. Did you think they'd be hotter? Well, they're pretty big, so their surface is a lot cooler than their core temperature. Some of the biggest blue stars have cores up around 40 million°C (104 million°F). How hot the core of a blue star is depends on its size. The bigger it is, the hotter it is, and the hotter it is, the faster it burns. The biggest, hottest stars burn the brightest, but only for a few million years. If they had a motto, it would have to be: live fast, burn hot, and die young.

As Light as Air

Q

Does air weigh anything? You can't tell by looking at it, and you can't feel it, but just because we can't see it or feel it, does it mean that air is weightless? You probably know that the answer is no. Air isn't heavy enough to register on your weight-sensing nerves, but it does have weight, and it's usually measured in cubic meters, or cubic feet or yards. So, how much could air possibly weigh? When different kinds of air are at the same temperature . . .

Which is the lightest?

a) dry air
b) moist air
c) very dry air
d) they all weigh the same amount

As Light as Air

Which is the lightest?

a) dry air
b) moist air
c) very dry air
d) they all weigh the same amount

CORRECT ANSWER:
b) moist air

It seems counter-intuitive, but moist air weighs roughly half as much as dry air. Dry air weighs about 1 kilogram (2.2 lb) and moist air weighs 0.6 kilograms (1.3 lb) per cubic metre (1.3 cubic yards) when both are at the same temperature. Air is made up of various components — mostly nitrogen, oxygen, and water vapour. Of the three, water vapour is the lightest. When there's more water vapour in the air, there's less oxygen and nitrogen, so the air weighs less. Most people think that because it's harder to breathe very moist air, it must be heavier, but it's not. It's just the humidity making it seem that way.

Q

It was three o'clock in the morning, and a young biologist, Ben Wilson, was alone in the lab with a fish tank holding 10 wild-caught herring. He was about to test the reaction of the fish to the sound of the hunting call of their mortal enemy, the killer whale. Seemingly out of nowhere, a loud rasping noise ripped through the lab. At first Wilson thought it must be a practical joke, so he searched the lab for the pranksters who were disrupting his experiment with rude noises. When he didn't find anyone, it dawned on him that it had to be the fish. He had just discovered that herring fart. It soon became apparent that in certain situations, they really let it rip.

When did the fish fart the most?

a) after eating
b) at night
c) when they were alone
d) when they were hungry

When did the fish fart the most?

a) after eating

b) at night

c) when they were alone

d) when they were hungry

CORRECT ANSWER:

b) at night

The fish farted the most at night, when the lab was quiet and dark. The rude noise was accompanied by a train of bubbles, ejected from their anal pores. They didn't fart after eating, and they didn't stop farting if they weren't fed. Regardless of their food intake, the herring farted up a storm at night, and the more fish in a tank, the more they all farted. What's up with that? The researchers think that it's social behaviour, and could be a form of communication. Wilson named the sound Fast Repetitive Tick, or FRT. Who says scientists don't have a sense of humour?

The Ping of Sonar

Q

Leonardo Da Vinci wrote about a tube that was lowered into the sea to listen for ships. About 400 years later, in 1917, French, British, and Canadian physicists put their heads together and pooled their knowledge to come up with a practical way to detect submarines. By 1918, they had an apparatus that could detect a submarine 457 metres (1,500 ft) away, and the famous "ping" sound you hear in old submarine movies rang out in the real world for the first time. Twenty years later, the technology was commonly referred to as sonar. Sonar is an acronym, a word made up of parts of other words.

What does sonar stand for?

a) Signal Over NAtural Rate
b) Sonic Output NAval Reflector
c) SOund Navigation And Ranging
d) Submarine Oceanic NAutical Radar

The Ping of Sonar

What does sonar stand for?

a) Signal Over NAtural Rate
b) Sonic Output NAval Reflector
c) SOund Navigation And Ranging
d) Submarine Oceanic NAutical Radar

CORRECT ANSWER:

c) SOund Navigation And Ranging

British scientists called it supersonics at first, but changed the name to ASDics to maintain secrecy. The ASD stood for Anti-Submarine Division. The word sonar was coined in America during World War II, and it's been with us ever since. Maybe because it's catchier than ASDics. The physicists who developed submarine-detecting technology had no idea that they were reinventing the wheel, so to speak. The ancient ancestors of today's toothed whales located submarine life forms (read: lunch) 40 million years ago with biosonar, or echolocation. Whatever you call it, it's nature's version of sonar, and it's still more accurate than manmade technology.

The Sound of Loud Music

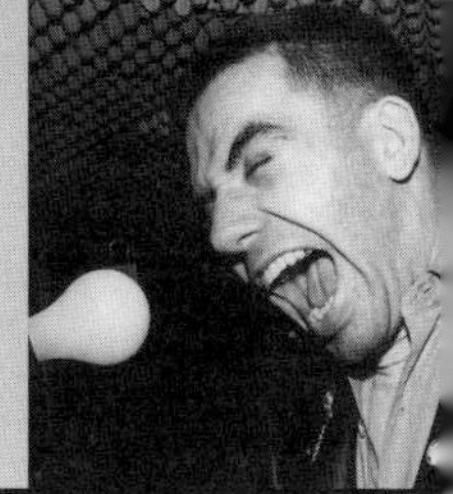

Loudness is measured in decibels, or dB. The range starts at 0 dB, the level of the quietest sound the human ear can detect. A soft whisper is 30 dB. Normal conversation is around 60 dB. A noisy restaurant might be 70 dB. Factories are often louder than 80 dB, and if the volume gets higher than 85 dB, ear protection is recommended. Most people set their personal listening devices louder than that. How much louder depends on where you live. In the European Union, the maximum volume allowed in portable music players is 100 dB, but elsewhere, volume is unrestricted. Let's say you're listening to your tunes at 100 dB . . .

How loud is your music? It's actually about as loud as . . .

a) a low-flying jet
b) operating a chain saw
c) someone shouting in your face
d) all of the above

Q

The Sound of Loud Music

How loud is your music? It's actually about as loud as . . .

a) a low-flying jet

b) operating a chain saw

c) someone shouting in your face

d) all of the above

CORRECT ANSWER:

d) all of the above

The chain saw, a low-flying jet, and someone shouting in your face are all around 100 dB. Many personal listening devices go up to 130 dB. That's more volume than standing in front of the speakers at a loud rock concert, and beyond the point at which sound becomes painful. As far as your ears are concerned, listening at 100 dB for 15 minutes is the same thing as being exposed to the sounds of a noisy factory, without ear protection, at 85 dB, for eight hours. Like a song, sound isn't just about volume, it also about duration. The longer you listen, the more it stresses your ears, even at what you might consider to be normal volume. Some experts estimate that one in seven people under the age of 19 already has some form of hearing loss from the sound of loud music.

The Sound of Snowflakes Falling

How likely is it that two different scientists, one in Canada and one in America, would, within a few years of one another, discover the sound of snowflakes falling? It's happened. What's more, their finds were weather-related happy accidents. The Canadian scientist set out to record winter rainfall with an underwater microphone on a lake in British Columbia. When it started to snow, he recorded the sound of falling snowflakes instead. The American scientists chased a freak snowfall to a motel in Mississippi where they lowered their microphones into the swimming pool. It turns out that it doesn't matter where they're recorded. When snowflakes hit the water, there's no mistaking the sound.

What is the sound of snowflakes falling into water?

a) crash like when you drop a bowling ball

b) fluff like hitting a soft pillow

c) plop like big drops of rain hitting the ground

d) screech like a fire engine's wail

The Sound of Snowflakes Falling

What is the sound of snowflakes falling into water?

a) crash like when you drop a bowling ball

b) fluff like hitting a soft pillow

c) plop like big drops of rain hitting the ground

d) screech like a fire engine's wail

CORRECT ANSWER:

d) screech like a fire engine's wail

It wasn't quite what the scientists had expected to hear. They'd anticipated a "plink" sound, and there was one, but it was followed by a screech resembling the wail of a fire engine passing by, and it all happened in less than one ten-thousandth of a second! A likely explanation for the screeching is that when the flakes hit the water, tiny air bubbles get trapped below the surface. The water's surface tension and pressure make the bubbles vibrate, and they emit a screeching sound before they pop. We can't hear the screeching of snowflakes without technology, and maybe that's a good thing. Who'd want to listen to the sound of tiny bubbles screeching like speeding fire engines for hours on end?

Q

Blue whales are the largest animals on the planet, and they may be the biggest creatures that ever lived. Adults can reach 30 metres (100 ft) in length, and can weigh more than any dinosaur discovered to date. Although you can't throw a blue whale on a fish scale, it's estimated that the biggest ones weigh more than 180 tonnes (200 tons). A big blue whale weighs more than 2,000 large men. Your heart is roughly the size of your clenched fist, but the blue whale's heart is the size of a small car. Their tongues are huge too, and weigh about as much as an adult elephant. What about their brains? How big could a blue whale's brain be?

A blue whale's brain weighs about as much as . . .

a) 1 medium-size red watermelon

b) 4 elephants

c) 7-seat family van

d) 1,000 human brains

The Blue Whale's Brain

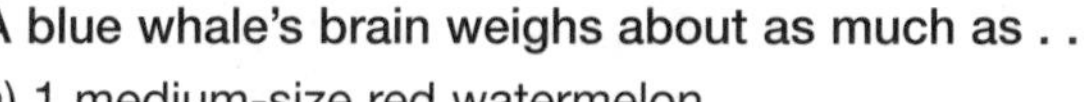

A blue whale's brain weighs about as much as . . .

a) 1 medium-size red watermelon

b) 4 elephants

c) 7-seat family van

d) 1,000 human brains

CORRECT ANSWER:

a) 1 medium-size red watermelon

Despite having a car-sized heart, the biggest animal on Earth has a brain that weighs as much as a medium-size red watermelon, 6–9 kilograms (13–20 lb). An elephant may weigh as much as a blue whale's tongue, but its brain, at 5.5–6.5 kilograms (12–14 lb), is in the weight range of blue whale brains. Elephants have the heaviest brains among land animals when it comes to absolute weight, but if you compare brain size to body size, humans are still number one. If blue whales' brains were as heavy as human brains relative to their body size, their brains would weigh about as much as a seven-seat family van.

The Whale's First Cousin

Every living thing on Earth is related, however distantly, because everything evolved from the same early life forms. As living things evolved, and became more complex, some creatures took surprising turns on their evolutionary path. In a way, cetaceans (whales and dolphins) made a U-turn. They were once aquatic creatures that developed into land animals, and then evolved into aquatic animals again. Up until fairly recently, there was a lot of speculation, but no proof of their mysterious and fascinating family history. It took genetic testing to reveal who the whale's closest living relative is.

To which animal are whales most closely related?

a) camel

b) cow

c) hippopotamus

d) wolf

The Whale's First Cousin

To which animal are whales most closely related?

a) camel

b) cow

c) hippopotamus

d) wolf

CORRECT ANSWER:

c) hippopotamus

You wouldn't think it by looking at them, but whales and hippos are first cousins. Their ancient ancestors were land animals that split into two groups about 50 to 60 million years ago. One group went back to the sea and evolved into cetaceans, and the other group stayed on land and evolved into water-loving, pig-like land animals. They were very successful for about 40 million years, but when they died out about 2.5 million years ago, they left just one descendent behind — the hippopotamus. Hippos are among the biggest and heaviest land animals, but even the biggest hippo is only about as heavy as a newborn blue whale calf. Hippos aren't completely aquatic, but they spend more time in the water than most mammals, not unlike the ancient ancestor they share with their first cousins, the whales.

The Whale That's a Whale, but Isn't

Q

Whales, dolphins, and porpoises all belong to the order *Cetacea*, which comes from the Latin *cetus*, which comes from the Greek *ketos*, meaning sea monster. Cetaceans are divided into baleen whales and toothed whales. Instead of teeth, baleen whales have plates made of keratin, the same stuff that hair and fingernails are made of, through which they filter small aquatic life forms. Toothed whales include dolphins, porpoises, and some whales. They have teeth, which they use to catch fish, squid, and other marine life. Did you know that some of the cetaceans we think of as being whales are actually oceanic dolphins?

Which species is NOT an ocean dolphin?

a) beaked whale
b) killer whale
c) melon-headed whale
d) pilot whale

The Whale That's a Whale, but Isn't

Which species is NOT an ocean dolphin?

a) beaked whale

b) killer whale

c) melon-headed whale

d) pilot whale

CORRECT ANSWER:

a) beaked whale

Beaked whales look like dolphins, but they're in a different family of the toothed whales. They're the deepest-diving air-breathing animals known. They've been observed going down almost 2 kilometres (1.2 miles), and can probably dive even deeper than that. Killer whales would more properly be called killer dolphins. They're the biggest of the ocean dolphins and were originally called whale killers, because they attack and kill other whales. The name got flipped at some point, and they became known as killer whales. Melon-headed whales and pilot whales are also large marine dolphins.

Big Melons

Toothed whales have a squishy, fleshy thing on their heads called a melon. If you've ever seen a beluga whale, you've seen one of the most magnificent melons of the sea. They're intelligent animals, so it's tempting to think that their melons are full of big, bulging brains. But, no, their melons are actually full of oily, fatty stuff. When a beluga, or other toothed whale, vocalizes or sends out echolocation signals, the flexible melon changes shape and acts like a lens to focus the sound into a beam. When an echolocation beam bounces back, it's intercepted. By what? Excellent question.

What intercepts a returning echolocation beam? The beluga's . . .

a) blowhole
b) ears
c) jaws
d) melon

Big Melons

A

What intercepts a returning echolocation beam? The beluga's . . .

a) blowhole
b) ears
c) jaws
d) melon

CORRECT ANSWER:
c) jaws

Toothed whales don't have external ears, but they do have a hollow lower jaw bone that's filled with fat, and it does for sound reception what the melon does for transmission. The fatty filling amplifies and conducts the returning sound signal and channels it directly to the middle ear, which sits inside the whale's head. The middle ear is insulated and protected by a layer of oil and air that prevents sound reception from any bone other than the jawbone. The returning echolocation signals from the middle ear are interpreted by the whale's brain to form a "picture" of its target object. You could say a toothed whale's lower jaw functions like ears, making it possible for toothed whales to "see" with sound.

A Little Echo and a Lot of Location

Echolocation evolved in toothed whales more than 30 million years ago, around the same time that their brains got bigger and their bodies got smaller. It was a big evolutionary leap forward. Dolphin echolocation is ten times more accurate than any manmade sonar system. Trained dolphins can pinpoint distant underwater mines with almost perfect accuracy. A blindfolded dolphin could detect something the size of a tennis ball a football field away. In experiments, at a distance of eight metres (26 ft), dolphins could sense which aluminum cylinder's walls were 0.23 millimetres (0.009 in) thicker than the rest. They echolocate by transmitting clicks lasting a fraction of a second and listening for the echoes, or reflected sounds. You know that the melon focuses the sound, but where does the echolocation click come from?

Where do echolocation clicks originate?
In the dolphin's . . .

a) fish lips
b) horse lips
c) monkey lips
d) none of the above

A Little Echo and a Lot of Location

Where do echolocation clicks originate?
In the dolphin's . . .

a) fish lips
b) horse lips
c) monkey lips
d) none of the above

CORRECT ANSWER:
c) monkey lips

Clicks are generated near the blowhole by the "monkey lips," the nickname biologists have for the dorsal bursa, or MLDB, complex. The monkey lips are a slit-like opening, and when air passes through them, they vibrate. Opening and closing the monkey lips creates a clack that makes the two fat-filled bursae near them vibrate, and they pass the sound wave to the melon. Are you wondering why they're called monkey lips? When dolphin anatomy was first studied, French scientists thought that the slit-like opening resembled monkey lips, and called it *museau de singe*, which means, literally, muzzle of the monkey. The French nickname was translated to monkey lips, and it stuck.

Q

It's not an urban legend! Of the toothbrushes tested by The Germinator, 10% were contaminated with fecal bacteria. It was one of the first of many puzzling discoveries made by The Germinator, a.k.a. microbiologist Chuck Gerba. His mission is to track down the baddest of the bad germs where they live. That's how he found them partying on toothbrushes. It's not as if people were cleaning the toilet with their toothbrushes, or rinsing their mouths with toilet water, so how was it happening?

Why were there fecal bacteria on the toothbrushes? Because . . .

a) people contaminated them with unwashed hands

b) people flushed the toilet with the lid up

c) the toothbrush holder was contaminated

d) they'd been dropped on the floor

Poopy Toothbrushes

A

Why were there fecal bacteria on the toothbrushes? Because . . .

a) people contaminated them with unwashed hands
b) people flushed the toilet with the lid up
c) the toothbrush holder was contaminated
d) they'd been dropped on the floor

CORRECT ANSWER:

b) people flushed the toilet with the lid up

Every toilet flush creates a plume of aerosols, tiny water droplets that carry bacteria and viruses up into the air. If the lid is down, it stops the plume from rising, but when the lid is up, the plume floats into the air and hangs around for at least two hours. That's more than enough time for the germs to settle on things nearby, including toothbrushes. To avoid a case of real potty mouth: store toothbrushes at least 1.8 metres (6 ft) away from the toilet, or inside a cabinet, and put the lid down before you flush. Problem solved. No more poopy toothbrushes.

Who's the Cleanest of Them All?

Q

You can't see them, but bacteria are all over the place. Most are harmless, but some are nasty pathogens that can make you very ill. Where are these creepy perps hiding out? The Germinator went into 15 average homes to track them down. He discovered all kinds of things, including who had the least germs at home.

Whose homes tended to be the least germy?

a) childless couples

b) families with children

c) single females

d) single males

Who's the Cleanest of Them All?

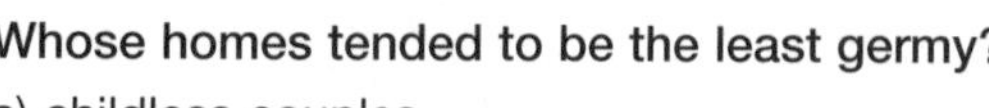

Whose homes tended to be the least germy?

a) childless couples

b) families with children

c) single females

d) single males

CORRECT ANSWER:

d) single males

Picture the stereotypical single male as seen in films and on TV: an unkempt-looking college student who never cleans or cooks, and only shovels out the mess in his place (maybe) when he's expecting female company. He'd rather do anything other than clean the bathroom. You'd think that single males would have the germiest homes of all. But Dr. Gerba found the opposite. Why? Let's see if you can figure it out.

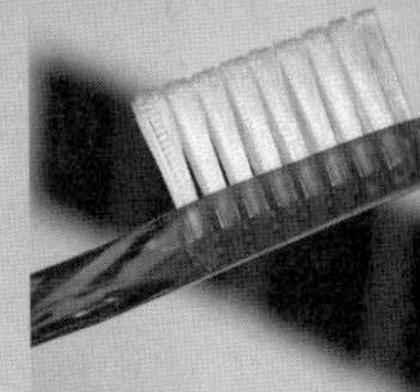

Swabbing the Poop Deck

Q

The Germinator had expected to find fecal bacteria in and around toilets in average homes, but he was surprised when they turned up in unexpected places in the bathrooms — on sinks, tubs, and countertops. Puzzling evidence, especially when you consider that bathrooms are often the most-cleaned area, except in the homes of single men, of course. Viruses and fecal coliform bacteria showed up in 80% of the bathrooms Dr. Gerba tested. Yet, according to an American survey, the average bathroom is cleaned about five times a month. So why were there nasty germs all over the place?

What was spreading the most fecal bacteria?

a) children

b) cleaning

c) toilet flushing

d) ventilation fans

Swabbing the Poop Deck

What was spreading the most fecal bacteria?

a) children
b) cleaning
c) toilet flushing
d) ventilation fans

CORRECT ANSWER:
b) cleaning

When the scientists watched people cleaning their bathrooms, the scary answer was obvious. Almost half used a rag or sponge to clean the toilet, and then clean something else, like the sink, tub, or counter. Instead of cleaning, they were actually spreading the crappy bacteria around. The Germinator thinks that this is why single men's homes had the fewest germs. They didn't clean often, so they didn't spread the crud around. That's one way to win the war on germs. The other way to win is by following The Germinator's battle plan: use paper towels and a disinfectant that kills both viruses and bacteria. Clean the least contaminated area first, and work up to the dirtiest jobs. Swab the poop deck, meaning the toilet, last. Toss the germy paper towel into the recycle bin. Congratulations. You've temporarily won the war on germs.

What Goes in the Wash . . .

Q

Once upon a time it took all day to do the laundry. Clothes were scrubbed with soap, boiled on the stove, wrung out by hand, and then hung out on the line to dry. These days, it's a lot less time consuming if you let machines do the dirty work for you. You toss the clothes into the washing machine, add detergent, and it does its thing while you do something else. Next comes the dryer, and it's done, all except for the sorting and folding. The clothes that come out of the dryer sure do look fresh and smell clean, but are they really? When The Germinator studied laundry . . .

What did he find?

a) Fecal bacteria can contaminate washing machines.

b) Sickening germs survive normal wash and dry cycles.

c) Underwear is the dirtiest of the dirty laundry.

d) all of the above

What did he find?

a) Fecal bacteria can contaminate washing machines.
b) Sickening germs survive normal wash and dry cycles.
c) Underwear is the dirtiest of the dirty laundry.
d) all of the above

CORRECT ANSWER:

d) all of the above

What goes in the wash stays in the wash if you don't kill them dead. Poopy bacteria can survive "normal" laundering, meaning cold or warm water washes with detergent, followed by a half hour in the dryer. Even worse, the germs liberated from underwear by detergents can survive quite nicely in the washing machine and contaminate all the loads that follow. Gross. Underwear needs to be washed in a separate load, in hot water, with bleach, followed by at least 45 minutes in a hot dryer. Colour-safe bleach will decontaminate coloured undies, but Dr. Gerba suggests using a bit more than the recommended amount. Better safe than poopy.

The 10 Most Contaminated Things in Your Home

Q

The Germinator tested 15 commonly used items or areas in homes in his quest to locate crappy germs such as E. coli and salmonella. In kitchens, his search for pathogens included dishcloths or sponges, sinks, sink faucet handles, counter tops, cutting boards, tables, and refrigerator handles. In bathrooms, he swabbed sinks and toilets. In the rest of the house, he tested floors and rugs, TV remotes, cell phones, the inside of washing machines, laundry hampers, and clothes. He found what he was looking for, in more than one place.

Where did he find the most fecal bacteria?

a) cutting boards

b) dishrag or sponge

c) toilet seat

d) TV remote

The 10 Most Contaminated Things in Your Home

Where did he find the most fecal bacteria?

a) cutting boards
b) dishrag or sponge
c) toilet seat
d) TV remote

CORRECT ANSWER:

b) dishrag or sponge

Surprised? So was Dr. Gerba. Sixty percent of the sponges harboured fecal bacteria and had, on average, a total bacterial count of about seven billion. That's close to the number of people on the planet — living in a dish rag! And if that's not gruesome enough, there were more fecal bacteria in the kitchens than in the toilets! More than half of the kitchen sinks were contaminated. They had more fecal bacteria than toilet seats, as did cutting boards (both wooden and plastic). The least contaminated, at the very bottom of the list, were — surprise! — toilet seats. Just think, by disinfecting the kitchen regularly, you can make it as clean as your toilet seat.

The 10 Most Contaminated Things in Your Home

1. kitchen dishcloth or sponge
2. kitchen sink
3. bathroom sink
4. kitchen sink faucet handle
5. cutting boards
6. refrigerator handles
7. TV remote
8. cell phones
9. inside the washing machine
10. laundry hamper

Q

You've been invited to a friend's birthday party. You're starving when you arrive, and dinner is still hours away. But, luckily, hot, yummy hors d'oeuvres are going around. You see mini quiche tartlets, sausage rolls, and jumbo garlic shrimp. You grab one, and practically inhale it. Mmmm, mmmm!

After eating the first one, you feel . . .

a) a little less hungry, but ready for more

b) hungrier than before you started eating

c) one was just enough to satisfy your hunger

d) too full to eat another bite

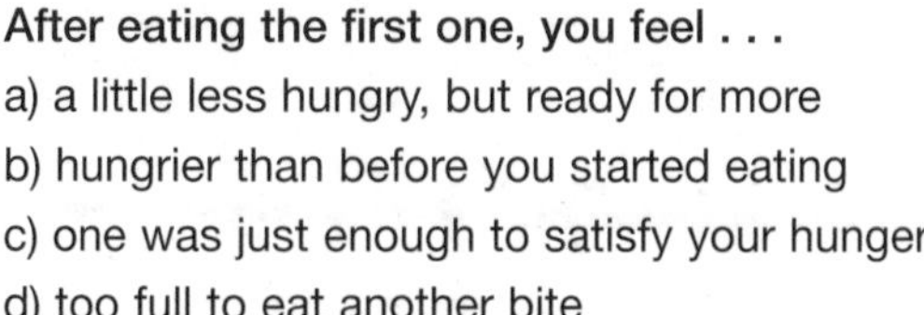

Love at First Bite

A

After eating the first one, you feel . . .

a) a little less hungry, but ready for more

b) hungrier than before you started eating

c) one was just enough to satisfy your hunger

d) too full to eat another bite

CORRECT ANSWER:

b) hungrier than before you started eating

It's counter-intuitive, but the aromas and initial bursts of flavour from that first yummy, little morsel make us even hungrier. But no problem, when you eat more appetizers, you eat less of the main course, right? Apparently not. In one experiment, people who were fed high-calorie soups before the main course consumed more calories during the main course than people who'd eaten low-calorie versions of the same soup. It would seem that when our appetite is turned on, we really go for the gusto.

Wine with a Healthy Kick

Q

You've heard that some of the compounds in wine are good for you and may even help you live longer. Some compounds in wine, called procyanidins, are supposed to be good for your cardiovascular system and might increase your longevity. According to researchers, you can receive the benefits from drinking just one glass of wine a day. But there's a catch. Procyanidins are only found in high amounts in certain wines. You check out the bottles set out on the bar.

Which wine should you choose?

a) French champagne

b) German white wine

c) Portuguese rosé

d) Sardinian red wine

Wine with a Healthy Kick

Which wine should you choose?

a) French champagne
b) German white wine
c) Portuguese rosé
d) Sardinian red wine

CORRECT ANSWER:
d) Sardinian red wine

When red wines were tested, Sardinian wines, and wines from the southwest of France, had the most procyanidins by far — sometimes as much as 10 times the amount found in wines from elsewhere. Why? They're made the old-fashioned way. Wines used to be fermented for three or four weeks to extract all the goodness from the skins and seeds, which contain the most beneficial compounds. These days, most winemakers don't ferment the grapes for much longer than a week, not enough time to extract the full health kick from the grapes. Some grapes pack more procyanidins than others; Cabernet Sauvignon and Nebbelio grapes tested the highest. You pour yourself a glass of robust Sardinian wine brimming with procyanidins.

Cheese, Please

Q

Glass of wine in hand, you turn your attention to a beautiful cheese tray by the wines and select a delectable cube of cheesy goodness. The relationship between wine and cheese is legendary, and they're paired so often, it's practically a marriage. Everyone knows that wine and cheese go together. That's probably why there have hardly been any scientific studies of their relationship until recently. When wine and cheese were taste tested together, what do you think the researchers found?

Cheese enhances the taste of . . .

a) all wine

b) cheap wine

c) red wine

d) white wine

Cheese, Please

Cheese enhances the taste of . . .

a) all wine
b) cheap wine
c) red wine
d) white wine

CORRECT ANSWER:
b) cheap wine

The taste of cheap wine was enhanced by cheese, but it turned out that cheese didn't enhance the taste of good wines, as people have long believed. It's the exact opposite, in fact. Cheese mutes the taste of wine. In the study, eight red wines and eight different cow's milk cheeses were tasted together in all possible combinations. The wines were perceived to be less oaky, less sour, less astringent, and with less berry flavour when tasted while eating cheese. The only flavour that was enhanced by cheese was butteriness, which was probably from the cheese and not from the wine. The muting of flavours might not be the best thing for appreciating the subtle nuances of a great wine, but it can definitely help a bottle of plonk taste better.

Good Friends, Good Times

Everyone's having a good time at the party. You've all known one another since way back when, and you all get along, so there's lots of talking, joking, and laughing. Time flies as the hors d'oeuvres, wine, and cheese are consumed, and before you know it, it's time for the festive, sumptuous dinner. You all sit down at the table together to dig in.

How much will you eat?

a) 22% less than normal
b) 22% more than normal
c) 44% more than normal
d) the same amount you normally eat

Good Friends, Good Times

How much will you eat?

a) 22% less than normal

b) 22% more than normal

c) 44% more than normal

d) the same amount you normally eat

CORRECT ANSWER:

c) 44% more than normal

If you're eating with people you know and love (or even just like), you can expect to pack away about 44% more food than you'd normally eat, maybe even more than 50% more. There's something about good company that revs up our consumption. The more people at the table, the more we eat. The more we're enjoying ourselves, the more we eat. The tastier the food is, the more we eat and chances are, we won't even be aware that we're overeating.

Winding Down with Caffeine

After dinner come the obligatory birthday cake and toast to the birthday girl, followed by more fun and laughs. But now it's starting to get late. You know that the party is winding down because your gracious hosts are serving coffee and tea. You could use a little caffeine, but not too much. You don't want to be up all night.

Which has the least caffeine? A . . .

a) cup of filtered coffee
b) cup of black tea
c) cup of green tea
d) shot of espresso

Winding Down with Caffeine

Which has the least caffeine? A . . .

a) cup of filtered coffee
b) cup of black tea
c) cup of green tea
d) shot of espresso

CORRECT ANSWER:

d) shot of espresso

How much caffeine there is depends on how much coffee or tea is used, what kind it is, how it's brewed, or how long it's steeped. In general, a shot of espresso has the least caffeine, about 40 milligrams. It seems counter-intuitive because it's strong, but espresso is served in very small quantities, usually only about 30 millilitres (1 oz). Both black and green tea have more caffeine, about 50 to 100 milligrams per 177-millilitre (6 oz) cup. A cup of filtered coffee has biggest hit, anywhere from 80 to 175 milligrams of caffeine. You knock back a shot of espresso, say your thanks and goodbyes, and step out into a starry moonlit night.

Q

The only earthlings who know what the moon smells like are the Apollo astronauts who walked on it. They couldn't exactly take their helmets off to sniff the surface directly, but they still managed to get a noseful. During every moonwalk, they got coated from head to toe in fine, sticky moondust. Once they were back inside their lunar module, when their helmets came off, they could smell the moon.

What does the moon smell like?

a) burnt gunpowder

b) green cheese

c) new paper money

d) wet dog

What Does the Moon Smell Like?

What does the moon smell like?

a) burnt gunpowder
b) green cheese
c) new paper money
d) wet dog

CORRECT ANSWER:
a) burnt gunpowder

All the astronauts who got a whiff of moondust described the smell the same way. According to Apollo 16 pilot Charlie Duke, not only did moondust smell like burnt gunpowder, it tasted like it too. That's odd, and not just because someone knows what burnt gunpowder *tastes* like. No, it's odd because moondust and gunpowder have almost nothing in common. Back on Earth, in the lab, the mystery deepened. The moondust didn't smell like much of anything at all. So, what's up with moondust? There are several theories, and lots of speculation, but no one knows for sure. Scientists will have to wait until the next series of moon missions to sniff out the secrets of the moon's intriguing scent.

Follow the Row of Craters and Hang a Right

Q

The first moon mission focused on science was Apollo 15. The crew was all Air Force pilots, but they studied geology for years, so that they could bring back the kinds of moon rocks that the scientists wanted to study. The astronauts pored over maps of their landing site, and burned the images of the nearby terrain onto their brains. The mental maps would help them to recognize targeted rock collecting areas once they were on the surface, but more importantly, it would make it possible to pilot the lunar module to the landing site. They had to follow a row of four craters that led to the spot. The last crater in the row was called Index.

What were the other three craters called?

a) Alpha, Beta, Gamma

b) Footnote, Glossary, Reference

c) Matthew, Mark, Luke

d) Pinkie, Ring, Middle

Follow the Row of Craters and Hang a Right

What were the other three craters called?

a) Alpha, Beta, Gamma
b) Footnote, Glossary, Reference
c) Matthew, Mark, Luke
d) Pinkie, Ring, Middle

CORRECT ANSWER:
c) Matthew, Mark, Luke

Matthew, Mark, Luke, and Index pointed the way to the landing site. The astronauts wanted to name the fourth crater John. NASA, however, had recently been sued by an atheist, after the Apollo 8 crew read a passage from Genesis while orbiting the Earth on Christmas Eve in 1968. Madalyn Murray O'Hair's lawsuit was eventually dismissed, but NASA opted to forego John, and go for Index instead. It made sense because Index Crater was closest to the landing site. Another crater visited by the astronauts, Spur, yielded one of the oldest moon rocks ever found. Back on Earth, it gave scientists a glimpse of what the solar system was like 4.5 billion years ago, when the moon was still forming. Ironically, it was dubbed "The Genesis Rock" by the media, and the name stuck — and no one sued.

High Lunacy

About 295 kilograms (650 lb) of the rocks, core samples, pebbles, sand, and dirt collected on the moon are locked away in a big vault at the Johnson Space Center. Even amateur felons would know not to target the big vault, but there was a safe in a lab at the space centre that proved to be irresistible to three student interns. The safe contained tiny samples from every Apollo mission. Ka-ching! The interns hatched a devious plan to steal the moon samples and sell them for $1 million. But before they stole anything, they cunningly tried to line up buyers by advertising "Priceless Moon Rocks Now Available!!!"

What happened after that?

a) A photographer tipped off the authorities.

b) They stole the moon rocks then tried to sell them to undercover agents.

c) They were busted on the 33rd anniversary of the first moon landing.

d) all of the above

High Lunacy

What happened after that?

a) A photographer tipped off the authorities.
b) They stole the moon rocks then tried to sell them to undercover agents.
c) They were busted on the 33rd anniversary of the first moon landing.
d) all of the above

CORRECT ANSWER:

d) all of the above

Posing as "Orb Robinson," one of the interns flogged the moon rocks to a mineralogy club in Belgium, where photographer/rock hound Axel Emmermann grew suspicious enough to contact the FBI. Undercover agents posed as buyers, and made a deal with "Orb." On the night of July 13, 2002, the interns entered the Johnson Space Center using their passcards, and broke into the lab. But since they couldn't open the 272-kilogram (600 lb) safe, they cleverly loaded it onto a cart, and wheeled it out of the building, and into a truck. A week later, on the 33rd anniversary of the first moon landing, the moon rocks deal was set to go down in an Italian restaurant in Orlando, Florida. But instead of driving off into the sunset with a wad of cash, the interns were driven to the police station with nothing but heartache. They were astronaut hopefuls who might even have made it into the space program. You have to wonder — what *were* they thinking?

The Great Ongoing Moon Experiment

An hour before the end of their final moonwalk, the first astronauts on the moon, Apollo 11's Neil Armstrong and Buzz Aldrin, set up an experiment in the Sea of Tranquility. It's the only experiment left on the moon by the Apollo missions that's still running today. From it, we've learned that the moon is spiralling away from the Earth. By how much? Well, let's say that the next moon landing mission is in 2019, and that when it's launched the Earth and moon are in the same positions relative to one another as they were during the Apollo 11 launch.

How much farther would the future astronauts have to fly?

a) 2.5 cm (1 in)
b) 2 m (6 ft)
c) 500 m (1640 ft)
d) 1 km (0.6 miles)

The Great Ongoing Moon Experiment

How much farther would the future astronauts have to fly?

a) 2.5 cm (1 in)
b) 2 m (6 ft)
c) 500 m (1640 ft)
d) 1 km (0.6 miles)

CORRECT ANSWER:

b) 2 m (6 ft)

Did you think it would be farther? The moon's average distance from Earth is increasing by about 3.8 centimetres (1.5 in) a year. So in the 50 years from 1969 to 2019, the moon would only be about 2 metres (6 ft) farther away. We know these things because of the four Lunar Laser Retroreflector Arrays left on the moon by the Apollo missions. Each array consists of a 60-centimetre (2 ft) wide panel holding 100 mirrors. Scientists on Earth ping the mirrors with laser telescopes. The mirrors were designed to send the pulse straight back to where it originated. The returning pulse is usually just a single photon, but it's enough to measure the moon's distance to within a few centimetres. So how far away is the moon anyway?

Are We There Yet?

Q

Gazing up at the moon, it's impossible to tell how far away it is. It took the Apollo astronauts about three days to get there — but then again, they were flying at an average speed of approximately 5,500 km/h (3,400 mph). Impressive, eh? But does it give you a sense of how far away the moon really is? Not me. I've never experienced spaceflight, but I can definitely relate to being in a car. How about taking the ultimate imaginary road trip: driving a car all the way to the moon! Let's say that the moon is suspended in a time warp, and standing still while the car, which defies the laws of gravity, barrels toward it at 120 km/h (74.5 mph). There's no stopping for food, or restroom breaks, or to sleep.

How long would it take to drive to the moon?

a) 3 weeks

b) 6 weeks

c) 2.5 months

d) 4.5 months

Are We There Yet?

How long would it take to drive to the moon?

a) 3 weeks
b) 6 weeks
c) 2.5 months
d) 4.5 months

CORRECT ANSWER:
d) 4.5 months

It would take about four and a half months of non-stop driving at 120 km/h (74.5 mph) to get to the moon on your imaginary road trip. By pinging the mirrors on the moon, scientists know that its average distance from Earth is about 385,000 kilometres (239,000 miles). If you want to do the math, divide the distance to the moon by the speed at which you're travelling. Whether you're calculating in kilometres or miles, it comes to about 3,208 hours travel time. Next, divide 3,208 by 24 hours, which comes to 133.6 travel days. That's about four and a half months. Think of how many sandwiches you'd have to make.

Q

There are about 240 million cars in the United States. That's enough to give one to every man woman and child in Canada, Mexico, Australia, New Zealand, and the U.K., as well as every person on all the islands of the Caribbean (excluding the tourists). America is one of the largest countries in the world, so even though there are almost as many cars as people, there are still a few parking spaces left. What about countries that aren't as big? When it comes to how many cars there are per square kilometre or mile in any given country, the U.S.A. is not number one in the world (for a change).

Which country has the most passenger cars per square kilometre or mile?

a) Germany
b) Italy
c) Japan
d) U.K.

Which country has the most passenger cars per square kilometre or mile?

a) Germany
b) Italy
c) Japan
d) U.K.

CORRECT ANSWER:

c) Japan

Japan was number one, with about 160 per square kilometre (256 per square mile), when researchers crunched the numbers for the density of cars in the G7 nations (Canada, France, Germany, Italy, Japan, the U.K., and the U.S.). Germany had the second highest density of cars, followed by Italy, the U.K., and France. The U.S. was second to last with about 25 cars per square kilometre (40 cars per square mile). Canada had the fewest cars by area, about 1.9 per square kilometre (3 per square mile). There were roughly 900 million cars on the planet in 2007. By now, there are probably 1,000 million, or more.

Symbols of Success

You probably know what these symbols mean. When we're familiar with them, symbols and logos are shorthand for our brains, and deliver a lot of information at a glance. A recognizable symbol, or logo, is money in the bank in the business world. Luxury items are almost always associated with memorable, classic logos. You may recognize the logos and link them to a brand, but do you know what the symbols stand for?

What does the Mercedes logo represent?

a) earth, wind, and fire

b) land, air, and water

c) peace

d) prosperity

Symbols of Success 

What does the Mercedes logo represent?

a) earth, wind, and fire

b) land, air, and water

c) peace

d) prosperity

CORRECT ANSWER:

b) land, air, and water

The Mercedes logo may be associated with prosperity in our minds, but that's not what it represents. Before his engines were successful, Gottlieb Daimler sent his wife a postcard on which he'd drawn a three-pointed star, and written that one day the star would "shine over our triumphant factories." Daimler believed that his engines should be used for land, air, and water, and that's what the three points of the star represent. After Daimler's death, when the Mercedes needed a trademark logo, his sons remembered the star and, in 1910, the symbol adorned the grille of a Mercedes for the first time. The logo has been updated over the years, but if you saw the very first one, you'd probably recognize it.

The World's Most Expensive Cars

From the beginning, the desire for greater speed and more luxury has driven the development of the automobile. Even back in the earliest days, there was a market for luxury vehicles. They were the most technically advanced cars and were built with the rich in mind, often to the buyer's specifications. At the 1921 Berlin Motor Show, a car that would be individually designed for every buyer debuted. No two would be exactly alike. In 1929, the same car maker introduced the first automobile with a V12 engine, and then in the early 1930s, a super-luxury 5.5-metre (18 ft) long version of this car was made.

What was this super-luxury car called?

a) Jaguar
b) Koenigsegg
c) Rolls-Royce
d) Zeppelin

The World's Most Expensive Cars

What was this super-luxury car called?

a) Jaguar
b) Koenigsegg
c) Rolls-Royce
d) Zeppelin

CORRECT ANSWER:

d) Zeppelin

Karl Maybach, the son of Daimler's partner, was building engines for Zeppelin airships and other aircraft in 1909. He was also making cutting edge car engines that he couldn't sell, so he decided to build the best cars of the era himself. He made fewer than 2,000 Maybach luxury cars in all, and of those, 183 were Zeppelins. Each one cost more than 30 regular cars. These days, a Maybach in good condition is worth anywhere from US$1–1.5 million. That's enough to buy 30 to 50 regular cars today. Maybach succeeded in building some of the best cars of his era, and some of the priciest collectible cars of all time.

Of the luxury cars made today, these are the ones with the biggest price tags, ranging from well over US$1 million to about half a million.

The Top Ten Most Expensive Cars in the World

1. Bugatti Veyron
2. Ferrari Enzo
3. Pagani Zonda C12F
4. SSC Ultimate Aero
5. Leblanc Mirabeau
6. Keonigsegg CCX
7. Saleen S7 Twin-Turbo
8. Koenigsegg CCR
9. Porsche Carrera GT
10. Mercedes SLR McLaren

Your Brain on Cell Phone

Q

There you are, driving your car and talking on your cell phone. Maybe you've heard that it's not a good idea to use it while you're behind the wheel, but you feel alert, the traffic is moving, and you have a hands-free cell phone, so, no problem, right? That's what American scientists wanted to find out when they ran a driving simulator experiment, where the subjects were tested while using, and while not using, cell phones. They compared the results and found that . . .

Talking on a cell phone while driving . . .

a) makes a 20-year-old's brain function like a 70-year-old's
b) slows reaction time by almost 20%
c) makes drivers twice as likely to be involved in a rear-end collision
d) all of the above

Your Brain on Cell Phone

Talking on a cell phone while driving . . .

a) makes a 20-year-old's brain function like a 70-year-old's

b) slows reaction time by almost 20%

c) makes drivers twice as likely to be involved in a rear-end collision

d) all of the above

CORRECT ANSWER:

d) all of the above

The brains of 20-year-olds are sharp, and their reaction times are as fast as they'll ever be, but when they drove while on a cell phone, their brains reacted like a 70-year-old's. So what happened to the 70-year-old drivers in the experiment? Did their brains react like 120-year-old's? Apparently not. The seniors in the simulators had half as many accidents as the young drivers. That could be because older drivers are more experienced or more cautious. Regardless of age, everyone's reaction time suffered. The reason for the downturn in driving performance is something called inattention blindness, which means that you can be looking right at something but not see it because your brain is preoccupied — like when you're on your cell phone.

The World's Smallest Car

Q

In 2005, American scientists built the world's smallest car. The nanocar is just slightly wider than a strand of DNA, which means that you could park more than 20,000 nanocars, side by side, across an average human head hair. It was built from the parts found inside a single molecule, and features a chassis, axles, and pivoting suspension. The nanocar rolls along on four wheels made of tiny, pure carbon spheres called buckyballs. It took the dedicated scientists eight long years to perfect the world's smallest car.

What took so long? Figuring out . . .

a) how to attach the wheels

b) how to make it run

c) how to make it that small

d) which molecules to use for parts

The World's Smallest Car

What took so long? Figuring out . . .

a) how to attach the wheels
b) how to make it run
c) how to make it that small
d) which molecules to use for parts

CORRECT ANSWER:

a) how to attach the wheels

After almost a decade of trial and error, the persistent scientists finally figured out how to attach the wheels without destroying the car, and they proudly rolled out their microscopic vehicle. What did they use for fuel? They didn't. They pulled it along with a non-optical microscope's probe. Why did they build the nanocar? So they could prove that molecular machines work, and move on to designing and building the world's smallest nanotruck capable of carrying a payload, and the world's smallest light-driven nanocar, of course. If you ever need an atom or molecule transported, call Rice University. The scientists there have the vehicles you need.

Q

Have you ever opened a can or jar of mixed nuts and noticed that the big nuts were on top, and the small nuts were on the bottom? You'd think that the biggest and heaviest nuts would sink to the bottom, not rise to the top. It's counterintuitive, but this nutty tendency is deep physics. Mixed nuts, and some other granular materials, sort themselves the same way under the same conditions. Maybe you know something about why bigger nuts float to the top and smaller nuts sink to the bottom.

What is this tendency called?

a) Almond Dilemma

b) Brazil Nut Effect

c) Cashew Conundrum

d) Peanut Problem

A Nutty Tendency

What is this tendency called?

a) Almond Dilemma
b) Brazil Nut Effect
c) Cashew Conundrum
d) Peanut Problem

CORRECT ANSWER:
b) Brazil Nut Effect

Physicists don't quite agree on how the Brazil Nut Effect works. Some think that as the cans make their way from the factory to the store and, finally, to your home, they're shaken and jostled, which creates small gaps between the nuts. The peanuts sink into the gaps, fill up the bottom, and push the Brazil nuts to the top. This is called kinetic sieving. Others think that the shaking and jostling moves the nuts up through the middle, across the top surface, and down the sides of the container. The largest nuts are too big to slide down, so they stay on top. This is called vibration-induced convection flow. Some experts think it's a bit of both. The Brazil Nut Effect has to be one of the deepest, darkest, mysteries of the universe, because even after 70 years of study, physicists still can't say exactly how it works.

Chemicals of Mass Deduction

Q

Have you heard of a chemical called acrylamide? Too much of it causes cancer and genetic mutations in lab animals, and it's not good for humans either. When Swedish scientists found acrylamide in the blood of workers who handled chemicals, they weren't surprised, but when it showed up in lots of other people who didn't work with chemicals, their alarm bells went off. How could this chemical be contaminating so many people? They devised an experiment that soon revealed the answer to the puzzling evidence.

What did they do? They . . .

a) analyzed the emissions of a disposable diaper factory

b) checked the soil in potato farmers' fields

c) fed fried chow to lab rats

d) tested tap water that had been in a glass overnight

What did they do? They . . .

a) analyzed the emissions of a disposable diaper factory
b) checked the soil in potato farmers' fields
c) fed fried chow to lab rats
d) tested tap water that had been in a glass overnight

CORRECT ANSWER:

c) fed fried chow to lab rats

The lab rats fed fried chow for two months had acrylamide levels 10 times higher than rats fed regular chow. The scientists had suspected that high cooking temperatures were the problem, and the rats proved it. Food scientists all over the world began to test various foods for acrylamide. Potato chips had the highest levels, followed by french fries, crisp breads, baked goods, and some cereals, but it was found in cooked meats too. Should you be worried? If you're eating a healthy diet, acrylamide-heavy foods are OK occasionally. If you prefer a 100% acrylamide-free diet, the options are raw, boiled, or steamed foods. Boiling water is hot, but not hot enough for acrylamide to form.

The Space Pen

Have you heard that NASA scientists spent 10 years and $12 million dollars developing a space pen that could write in zero gravity, underwater, on most surfaces, and in extreme temperatures? The story often ends with the punchline "and the Russians used a pencil." The story is, of course, an urban legend. The truth is that the space pen was invented by Paul Fisher, a pen-maker, at his own expense. NASA tested it for two years and paid just $2.95 per pen. First used in space on Apollo 7 in 1968, they were the first ballpoint pens that could actually write in zero gravity. Among the technological innovations that made the space pen a success was the ink.

What flows like the ink in the space pen?

a) ketchup

b) lava

c) mercury

d) motor oil

The Space Pen

What flows like the ink in the space pen?

a) ketchup

b) lava

c) mercury

d) motor oil

CORRECT ANSWER:

a) ketchup

The ink in the space pen has a property called thixotropy, and it behaves a lot like thick ketchup that doesn't want to flow out of the bottle. One way to get the ketchup going is to rap the end of the bottle. The force of the blow temporarily makes the ketchup runny, and it flows more easily. The ink works in a similar way. It's inside a pressurized cartridge that has a plug of nitrogen gas. The gas pressure feeds the ink to the ball in the point, which shears the ink when it rolls. The shearing action liquifies the thixotropic ink, and it flows like ketchup flows out of a bottle that's just been rapped.

The World's Lightest Solid

Q

What's 99.8% empty space, and weighs just a little more than the air you breathe? It's 1,000 times less dense than glass, and very spongy but, in space, it easily absorbed the impact of dust particles shearing into it at six times the speed of a rifle bullet. Have you figured out what this remarkable material is? It's the world lightest solid: aerogel. You might think this space age material is a recent innovation, but the great grandaddy of all aerogels was invented in 1931.

What was aerogel first used for?

a) fashion jewellery
b) picnic cooler insulation
c) thickening agent in napalm bombs
d) water filtration

The World's Lightest Solid

What was aerogel first used for?

a) fashion jewellery
b) picnic cooler insulation
c) thickening agent in napalm bombs
d) water filtration

CORRECT ANSWERS:

b) picnic cooler insulation, and c) thickening agent in napalm bombs

If you picked both correct answers, give yourself a virtual bonus point. After aerogel was invented, it was used for a few things, but it didn't really take off until space scientists got involved. They modified and improved the silicon-based material over the years, and made it incredibly lightweight, just 1.9 milligrams per cubic centimetre (0.06 cubic inches). Aerogel has gone from insulating picnic coolers to insulating the machines we send to explore other worlds, like Mars, not to mention the solar system. It was used to capture interplanetary stardust, and dust from the tail of Comet Wild 2. The dust turned out to be from about 4.5 billion years ago, when the solar system was forming. It may look like pale blue sponge toffee, but aerogel has provided scientists with insights into how our solar system, and the objects in it, formed.

Packaging Pollution

Who in the world isn't aware of the problems with the environment? Air, water, and land pollution have touched every corner of the globe, and then there's garbage. Most countries have more than they know what to do with, and much of that trash is packaging. Cutting down, or eliminating, packaging is the no-brainer solution to reducing the mountains of waste we generate. Most of the packaging materials we use are not truly biodegradable, but there are some. If you were to bury it . . .

Which packaging material would break down completely in one to two months?

a) cellophane

b) plastic

c) styrofoam

d) vinyl

Packaging Pollution

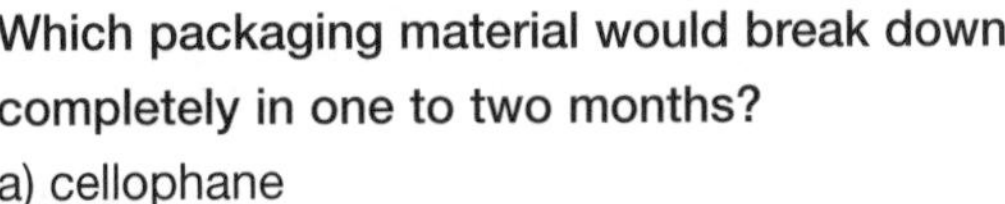

Which packaging material would break down completely in one to two months?

a) cellophane
b) plastic
c) styrofoam
d) vinyl

CORRECT ANSWER:

a) cellophane

Cellophane is made of cellulose fibres from wood, cotton, or hemp, and it breaks down faster than most paper. Uncoated cellophane film breaks down in a month or two, while the coated kinds take about twice as long. In lake water, cellulose breaks down even faster — about 10 days for uncoated, and a month for coated film. Even though it's completely biodegradable, cellophane still has a downside. The process used to make it creates pollutants. Budding inventors, this is your challenge: come up with a biodegradable packaging material with an earth-friendly manufacturing process.

Q

The world's first electronic instrument was invented by a Russian physicist and professional musician, in 1918. He called it the etherphone because the sounds it made were unearthly. He started out touring the Soviet Union giving etherphone music concerts, and it wasn't long before he was wowing the crowds at sold-out shows in Europe. Maybe America wasn't ready for electronic music in 1928, because when he made his concert debut in New York City, some people walked out. Others loved the etherphone experience and raved about it. Albert Einstein, who was a scientist and violin player, went to a concert and was intrigued.

How was the etherphone played? By . . .

a) breath control

b) moving levers up and down

c) turning knobs

d) waving the hands through the air

The World's First Electronic Instrument

How was the etherphone played? By . . .

a) breath control

b) moving levers up and down

c) turning knobs

d) waving the hands through the air

CORRECT ANSWER:

d) waving the hands through the air

Léon Theremin wanted to pull music straight out of the air in the same way as orchestra conductors, who wave their hands around to control the output of the instruments. Once Theremin started waving his hands around on American stages, the etherphone was identified with him and became known as the Theremin. It's the only instrument that's played without being touched by the player. Musicians wave their hands around two antennas attached to a small box that generates radio waves. Their movements interfere with the radio waves, and that interference produces the sounds. Players control pitch with a hand over one antenna, and volume with a hand over the other. The way the theremin is played looks like magic, but the ethereal sounds are firmly grounded in science.

How about a little time travel back to 1931? Your favourite big band is playing live. You can see the guitarist's hand moving rhythmically, but you can't hear the guitar over the horns and strings. No problem, you say? Just crank up the volume on the guitar? Well, you can't. Electric guitars, as we know them, haven't been invented yet. Early "electrified" guitars had hollow bodies, which made for feedback and resonance problems when players cranked up the volume. The electric guitar needed to be reinvented. Enter Les Paul, guitarist and inventor. He knew that Thomas Edison had once made a solid body violin, and he thought a solid body would fix the problems. So, he spent his Sundays working on it, and by 1941, he'd made the first solid body guitar with electric pickups.

What did he call it?

a) Blackie

b) The Broadcaster

c) The Les Paul

d) The Log

Crank It Up to 11!

What did he call it?

a) Blackie

b) The Broadcaster

c) The Les Paul

d) The Log

CORRECT ANSWER:

d) The Log

Les Paul made The Log out of a 10 centimetre by 10 centimetre (4 in by 4 in) block of pine, some phone parts (for the electric pickup), and the neck and bridge of a guitar. The first time he played it at a gig, people were so distracted by its weirdness, they couldn't focus on his music. So, he cut a hollow body Epiphone guitar in half lengthwise, and sandwiched The Log between the two halves. Thanks to the cosmetic makeover, The Log didn't distract from his music anymore, and the audiences raved about his sound. Leo Fender was also developing a solid body around the same time, and managed to get his Broadcaster on the market ahead of the Les Paul, the guitar based on The Log. Finally, guitarists could crank up the volume to the max.

Musical Objects of Desire

Adventurous musical artists are the force that drives music into the future. Pushing the envelope as far as they can, they explore the creative, and technological, possibilities of musical expression. When they work with brilliant scientists, they can alter the sound of popular music, or even revolutionize it. During a stellar musical explosion in the 1960s, some artists started to use a new technology, available for the first time. It's a common instrument now, but in the late '60s, it was expensive, hard to get, and at the forefront of a musical revolution. With that in mind . . .

Which recording does not belong?

a) Bee Gees' *Ideas*

b) The Beatles' *Abbey Road*

c) The Monkees' *Pisces, Aquarius, Capricorn & Jones*

d) Walter (Wendy) Carlos' *Switched-On Bach*

Musical Objects of Desire

Which recording does not belong?

a) Bee Gees' *Ideas*

b) The Beatles' *Abbey Road*

c) The Monkees' *Pisces, Aquarius, Capricorn & Jones*

d) Walter (Wendy) Carlos' *Switched-On Bach*

CORRECT ANSWER:

a) Bee Gees' *Ideas*

W. Carlos' *Switched-On Bach* blazed the trail as the first music album played entirely on the first keyboard-triggered music synthesizer, the Moog (rhymes with vogue). Carlos worked closely with the synth's inventor, physicist Robert Moog, to make it more musician-friendly. Moog wasn't a musician himself, but enjoyed hanging out with them, and developing instruments for them. His company made theremins, and they inspired the synthesizer's sound generating components. The Monkees and The Beatles used Moogs in the late 1960s, but not the Bee Gees (yet). In 1965, a Moog cost about US$10,000 — that's around $65,000 today! We take synthesizers for granted now, but back then, they truly were musical objects of desire.

Music from Space

Q

Not only do astronauts listen to music while they're in space, many of them make music too. After a long, hard day's work, astronaut-musicians like to unwind with their favourite instrument. Ed Lu played his keyboard on the International Space Station. Chris Hadfield played his collapsible guitar on a space shuttle mission, and then gave it to Thomas Reiter on the Space Station Mir. These are just a few relatively recent examples, but there's been music in space since the early days of spaceflight. All of the musical instruments listed below have been played at some time or other, but . . .

Which was the first instrument played in space?

a) didgeridoo
b) flute
c) harmonica
d) kazoo

Music from Space

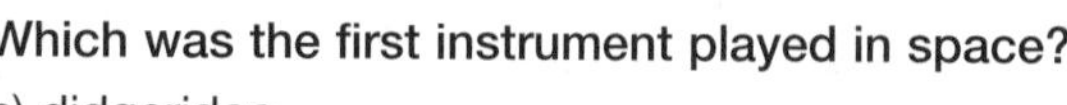

Which was the first instrument played in space?

a) didgeridoo

b) flute

c) harmonica

d) kazoo

CORRECT ANSWER:

c) harmonica

Gemini 6 and Gemini 7 were the first craft to rendezvous while in orbit around the Earth, in mid-December 1965. To celebrate, Wally Schirra and Thomas Stafford (Gemini 6) performed the first song in space, "Jingle Bells." Schirra played a tiny four-hole, eight-note Little Lady Hohner harmonica, and Stafford jangled a string of miniature sleigh bells. Now, if that's not spacy-sounding enough for you, Ken Bowersox and Don Pettit performed "Jingle Bells" on harmonica and didgeridoo on the International Space Station in 2002. The didgeridoo is an Australian Aboriginal wind instrument with a unique, haunting sound. Imagine it accompanying a harmonica. It wasn't the first version of "Jingle Bells" from orbit, and it probably won't be the last, but it has to be the most unusual duet ever played in space.

The World's Smallest Guitar

Q

Guitar collectors are an extravagant bunch. One collector was willing to pay close to US$1 million for guitar virtuoso Eric Clapton's old Fender Stratocaster, "Blackie." But there's one guitar that even the world's most avid collectors, or guitar legends, can't touch. It's the world's smallest guitar, and it was made by physicists at Cornell University in the U.S. How small is it, exactly? Here's a hint: it's smaller than a bread box.

How small is the world's smallest guitar?
About the size of . . .

a) a single cell
b) a poppy seed
c) the world's biggest bacterium
d) this period.

The World's Smallest Guitar

How small is the world's smallest guitar?
About the size of . . .

a) a single cell
b) a poppy seed
c) the world's biggest bacterium
d) this period.

CORRECT ANSWER:
a) a single cell

The world's smallest guitar was made for fun, and to show that mechanical devices can be made on a microscopic scale. The nanoguitar is about as long as a single cell, 10 micrometres (10 millionths of a metre, or about one 2,500th of an inch). It's anatomically correct, with six unimaginably thin strings, and, believe it or not, it's technically possible to play the itty bitty guitar. All you need to pluck the strings is a special microscope that can move atoms around. Even if you went to all the trouble, the guitar is so tiny you wouldn't hear it anyway. Isn't it time someone invented a nanoscale guitar pickup? The world's smallest electric guitar — now there's a fun challenge for today's musical scientists!

Q

The Allies used sneaky, secret weapons in World War II, and espionage was one of the big guns. Spies who could live by their wits under any circumstances were in high demand. To train more operatives for the war effort, North America's first spy school was set up. Silent killing, parachute drops, withstanding torture, deception, sabotage, and everything you need to know about plastic explosives were among the subjects on the curriculum. As many as 2,000 agents trained at this top-secret facility from 1941–44, including some who would later form the core of the CIA. Its location isn't top secret these days, so you might know . . .

Where was North America's first training school for spies located?

a) in Guanajuato, Mexico

b) in the suburbs of Rochester, New York

c) near Toronto, Canada

d) on Midway Island, South Pacific

Where was North America's first training school for spies located?

a) in Guanajuato, Mexico

b) in the suburbs of Rochester, New York

c) near Toronto, Canada

d) on Midway Island, South Pacific

CORRECT ANSWER:

c) near Toronto, Canada

The school, called Camp X, was located on a farm just east of Toronto, where some of the spies were recruited. According to a Camp X report, among the best spies were recent immigrants from countries with a (then) low standard of living, such as Yugoslavia and Hungary, as well as left-wing Spaniards and Italians. Most returned to their homelands to join the resistance against the Nazis. Liberal-minded Canadians, and sophisticates with dual nationality, preferably French and American, were also actively recruited. The Camp X Spying 101 course was 10 weeks long, but there were also weekend courses for intelligence officers. If I had to guess, I'd say that the future CIA directors probably just took the weekend courses.

The Real James Bond

Q

The mastermind behind Camp X became a master spy, but he started his illustrious career as a decorated World War I war hero. Next he became a successful inventor and businessman. Among his inventions is the great grandaddy of the fax machine — a wireless photo transmitter that made him a millionaire by the time he was 30. Officially, he was the British Passport Control Officer headquartered in New York City, but he was actually the head of BSC (British Security Co-ordination). His mission was to create a secret intelligence network that spanned the entire Western hemisphere. If you were sending him a top-secret coded message . . .

How would you address the telegraph? To:

a) 007

b) Deep Throat

c) Intrepid

d) Mother

The Real James Bond

How would you address the telegraph? To:

a) 007

b) Deep Throat

c) Intrepid

d) Mother

CORRECT ANSWER:

c) Intrepid

Years before World War II started, Canadian William Stephenson was living in England, but doing business in Germany regularly. When he noticed that Hitler's Nazi government was gearing up for war, he warned Winston Churchill. He volunteered to assassinate Hitler, but Churchill assigned him to head up British intelligence in the Western hemisphere instead. The telegraphic address of his headquarters was Intrepid, and it was later popularized as his code name. After the first James Bond movie, *Dr. No*, was released in 1962, Ian Fleming, who created the Bond character, wrote in *The Times*, "James Bond is a highly romanticized version of a true spy. The real thing is . . . William Stephenson."

What happens to intelligence officers when the war ends? Well, Sir William Stephenson went from being head of British intelligence in the Western hemisphere to being a successful businessman again, this time based in Bermuda. Some of the other officers who spent time at Camp X became prolific writers. They went from composing top-secret reports to penning unclassified, and extremely popular, fiction.

What was NOT written by a former intelligence officer?

a) *Beneath the Planet of the Apes* screenplay
b) *Charlie and the Chocolate Factory*
c) *Charlotte's Web*
d) *Chitty Chitty Bang Bang*

What was NOT written by a former intelligence officer?

a) *Beneath the Planet of the Apes* screenplay
b) *Charlie and the Chocolate Factory*
c) *Charlotte's Web*
d) *Chitty Chitty Bang Bang*

CORRECT ANSWER:
c) *Charlotte's Web*

Paul Dehn, Ian Fleming's friend and intelligence co-worker, wrote screenplays, including several *Planet of the Apes* sequels, *Goldfinger*, *The Spy Who Came in From the Cold*, and *Murder on the Orient Express*. Roald Dahl worked on the manuscript of the final Camp X report, and went on to write many children's books, including *Charlie and the Chocolate Factory*, as well as novels, books of poetry, short stories, and screenplays. Ian Fleming, who wrote dozens of novels and short stories about James Bond, the most famous fictional spy of all time, also wrote *Chitty Chitty Bang Bang* for his son, Caspar. *Charlotte's Web* was not written by a former intelligence agent, although author E.B. White was known to spy on the animals on his farm for inspiration.

The Biggest Aspidistra in the World

Q

Aspidistra was once the most powerful radio transmitter on the planet. At a time when most radio stations were 50 kilowatts or less, Aspidistra's transmitter was 600 kilowatts. Named for a popular wartime song about an Aspidistra flower so tall that the top of it was in the clouds, the 14-tonne (30,000 lb) top-secret monster transmitter was located in Crowborough, England. During World War II, it communicated with other top-secret Allied radio stations, including Hydra, the communications centre for British intelligence at Camp X. Both radio transmitters played important roles on the intelligence front. Hydra's job of receiving and retransmitting intelligence was straightforward. Aspidistra's role was more complicated, and you could say that Aspidistra played both sides.

What kind of radio station was Aspidistra?

a) black
b) grey
c) red
d) white

The Biggest Aspidistra in the World

What kind of radio station was Aspidistra?

a) black

b) grey

c) red

d) white

CORRECT ANSWER:

a) black

Aspidistra was a black radio station, meaning that it transmitted from England, but gave the impression that it was transmitting from within Germany. White stations broadcasted propaganda too, but didn't disguise their location. Black stations were first called Freedom Stations, and later, Research Units, or RUs. Whatever they were called, their purpose was psychological warfare and sabotage. Most of the information Aspidistra transmitted was from legitimate German news sources, but the real news was interspersed with propaganda designed to turn German citizens against their own leaders. Aspidistra was also used to derail bombing missions by confusing Luftwaffe pilots. Their coded instructions were recorded in the evening, and Aspidistra retransmitted the same instructions to the pilots on the next night's bombing run. More than the fighter jets were scrambled when that happened.

"The Thing" in the Ambassador's Office

A large, carved, wooden plaque of the Great Seal of the United States was presented to the U.S. Ambassador in Moscow by Soviet children as a gesture of friendship. It hung on the wall of the ambassador's home office, right behind his desk, from 1946 to 1952. It was only removed when a British radio operator discovered that there was a bug in the gesture of friendship. When he tuned in to an open channel, he could hear the conversations taking place in the Ambassador's office. A bug was found inside the plaque, right under the Eagle's beak. Rumour has it, it took intelligence and counterintelligence agents six months to figure out how "The Thing" worked. No one had seen anything like it before.

What made The Thing different from other spyware of the day?

a) It had no power supply or active electronic components.
b) It resembled a burglar alarm.
c) It was activated by a motion detector.
d) all of the above

“The Thing” in the Ambassador’s Office

What made The Thing different from other spyware of the day?

a) It had no power supply or active electronic components.
b) It resembled a burglar alarm.
c) It was activated by a motion detector.
d) all of the above

CORRECT ANSWER:

a) It had no power supply or active electronic components. (All of the above is not always the correct answer!)

The Thing was the first passive resonant cavity bug, and it consisted of nothing more than a tiny microphone connected to a small antenna. It radiated no signal when it was inactive, and worked on the same principle as the theremin electronic music instrument — radio wave interference. That’s no coincidence, because both were invented by Léon Theremin. When microwaves were beamed at The Thing, it acted like a tiny radio station, and transmitted the sounds it picked up in the Ambassador’s office to a receiver in another building. Once the Brits and Yanks had figured it out, bugs based on The Thing’s design were built into all kinds of things, including the walls of embassies. There, the ingenious little devices listened in, undetected, to countless conversations for years, if not decades.

Q

Four students are sitting a table, dutifully eating bowls of tomato soup. They think they're taking part in a tomato soup taste test, but they're not. They don't know that 1) it's not really a taste test, and 2) two of the four bowls on the table are bottomless, rigged to refill at the same rate at which the soup is being eaten. Sounds like a scene from a candid camera–type show, doesn't it? It could be, but it's actually a scene from a classic Brian Wansink food psychology experiment. He wanted to find out what tells people to stop eating: feeling full or visual cues.

What happened during the tomato soup experiment?

a) All the students ate the same amount.

b) Those eating from bottomless bowls ate more.

c) Those eating from bottomless bowls ate less.

d) Those eating from normal bowls ate more.

Bottomless Bowls

What happened during the tomato soup experiment?

a) All the students ate the same amount.
b) Those eating from bottomless bowls ate more.
c) Those eating from bottomless bowls ate less.
d) Those eating from normal bowls ate more.

CORRECT ANSWER:

b) Those eating from bottomless bowls ate more.

The students with the bottomless bowls ate 65% more, on average, than the students with the normal bowls. Some ate three times as much, including one unsuspecting student who slurped down more than a litre (quart) of tinned tomato soup. When Dr. Wansink asked him what he was doing, the student said that he was trying to get to the bottom of the bowl. No one noticed the bottomless bowls refilling, or wondered why the level wasn't going down. They just kept eating. When it comes to how much we consume, we tend to use our eyes, not our stomachs, to tell us when we're full.

If you were participating in a test, and were asked to choose the tastiest beverage, or the best quality T-shirt, you wouldn't just choose one at random. You'd decide after carefully considering the evidence, right? That's what most people think they do. To study the workings of our decision-making process, psychologists went to a mall and set up tables. They laid out four pairs of identical pantyhose labelled A, B, C, and D, and invited shoppers to choose the pair that was the best quality. The sneaky scientists didn't tell the unsuspecting participants that all the pantyhose were identical. Since they were all the same, you'd think that each pair would have been chosen about a quarter of the time, but that's not what happened.

Which pantyhose was chosen the least often?

a) A

b) B

c) C

d) D

Decisions, Decisions

Which pantyhose was chosen the least often?

a) A
b) B
c) C
d) D

CORRECT ANSWER:

a) A

Even though all the pantyhose were identical, A was picked the least often, with only 12% of the vote. B got more votes than A, and C got more votes than B. D was the most popular, with 40% of the participants vouching for its superior quality. The participants had very specific reasons for choosing one pair of pantyhose over the others. When the experimenters revealed that the pantyhose were identical, all the participants stuck with their rationale, convinced that their choice truly was the best quality. Why was A chosen the least, and D the most often? Studies show that when we're presented with objects placed in a row, we tend to choose the ones that are farther to the right. We do choose things for a reason, but maybe not for the reason we think.

Cold Weather Personalities

Q

Northerners who live where the winters are harsh are often described as being hardy. The implication is that people who live in cold climates are tougher somehow. Is there any truth to that? To find out if personality is a factor in cold tolerance, scientists at Canada's Defence Department tested, and rated, 20 subjects for five personality traits, and then subjected them to extreme temperatures changes. The participants sat in shorts and shirts in 10°C (50°F) for 90 minutes, and then in 40°C (104°F) for the same amount of time. Next, they were bundled up in warm clothes and exposed to the wind-chill equivalent of about -3°C (26°F). That's pretty cold, but not unusual for winter in some parts of the world.

Which personality type was most comfortable in the extreme cold?

a) adventurous

b) conscientious

c) extroverted

d) neurotic

Cold Weather Personalities

Which personality type was most comfortable in the extreme cold?

a) adventurous
b) conscientious
c) extroverted
d) neurotic

CORRECT ANSWER:
d) neurotic

During the cool and hot parts of the experiment, the neurotics complained more than the extroverts, and their bodies registered more stress. But in the extreme cold, the neurotics were more comfortable, and the extroverts were more stressed. The researchers hadn't expected the sudden role reversal, but speculate that maybe the extroverts, who were more comfortable in the first part of the experiment, were shocked by the unpleasant turn of events. The neurotics, meanwhile, had expected something bad to happen, and when it finally did, it wasn't a big deal. It's no secret that different people respond differently to temperature stress, but how, or why, is still a mystery.

The $ecrets of $uccess

Q

There are about a million millionaires in the world, which means that your odds of becoming one are better if you're in business than if you're playing most lotteries. Of course, buying a lottery ticket is a lot easier than succeeding in business. Aside from hard work, other things go into a winning formula for success, including an education, timely ideas, good connections, and luck, to name a few. Those at the bottom of the corporate ladder are always looking for an edge, something special, that will help them rise to the top. Is there a secret edge? According to studies . . .

What percentage of male executives are taller than average?

a) 15%
b) 30%
c) 60%
d) 90%

The $ecrets of $uccess

What percentage of male executives are taller than average?

a) 15%
b) 30%
c) 60%
d) 90%

CORRECT ANSWER:
d) 90%

The average height for North American males is 175 centimetres (5' 9"), but in studies, 90% of business executives were taller than average. Less than 4% of men in the general population are over 188 centimetres (6' 2"), but nearly a third of the execs were that tall. The "height bonus" for male execs is about $1,000 a year for every 2.5 centimetres (1 in) of height over 168 centimetres (5' 6"). For women, height wasn't a factor, but there was a premium for beauty, and a penalty for plainness. In one study, obese white women (but not black women) were heavily penalized. They earned 17% less than women in their recommended weight range. Attractive women earned about 5–10% more than average lookers. So, guys, if you want to be a successful CEO, be as tall as possible. Girls, you have to be as naturally, or surgically, attractive as possible. But sadly, even if you're attractive, in most of the working world, you're still not likely to earn as much as a guy.

Criminal Makeover

Q

Are unattractive men more likely to be criminals? Do good-looking criminals get better treatment? To study the relationship between attractiveness and criminality, 100 unattractive perps were given plastic surgery to improve their looks before they were released from a major American prison. They were matched with 100 equally unattractive men who didn't receive a cosmetic makeover before being released. A year later . . .

What did the researchers find?

a) The makeover ex-cons had more re-arrests.
b) The makeover ex-cons had fewer re-arrests.
c) The untreated ex-cons had fewer re-arrests.
d) They all fared about the same.

Criminal Makeover

What did the researchers find?

a) The makeover ex-cons had more re-arrests.
b) The makeover ex-cons had fewer re-arrests.
c) The untreated ex-cons had fewer re-arrests.
d) They all fared about the same.

CORRECT ANSWER:

b) The makeover ex-cons had fewer re-arrests.

The inmates who'd had procedures to correct broken noses or protruding ears, or who'd had ugly tattoos removed before they were released, did significantly better than the unattractive ex-cons. Why? Could it be that being better-looking makes you a better person? Or is it that attractive people are treated better, and get away with more? Well, studies leave little doubt that attractive people get preferential treatment in all walks of life. The bottom line is that no one knows exactly why the surgery worked, but if this study is any indication, it might pay to give cosmetic makeovers to criminals before they're released back into society. Better-looking ex-cons who are less likely to re-offend could prove to be a win-win situation for everyone.

EPILOGUE

Well, that's it for now. It's not that I've run out of questions, we've simply run out of pages in this book. So how did you do? If you aced the quizzes, congratulations! But even if you didn't, you're already smarter, simply because you challenged your brain and learned something new. Your brain may even have forged new neural pathways in the process.

This book was cooked up at The Brain Café™. It took shape in my imagination as a virtual café, symbolizing a relaxed but stimulated state of mind, a place where learning is pure pleasure. The Brain Café™ concept is expanding like the early universe and, at this point in time, includes *What Does the Moon Smell Like?* and a growing site on the world wide web where you'll find more to intrigue, entertain, and enhance your brain. The Brain Café™ is always open, never closed. Consider yourself invited!

Meet me @ The Brain Café™
www.thebraincafe.ca
Eva Everything

Image Credits

"A Whale We Go" image copyright Michael Price, iStockphoto; "By the Billion" image copyright Wendell Franks, iStockphoto; "Cellular Beings" image copyright Martin Fischer, iStockphoto; "Extremes" image copyright Robert Gendler; "Laugh Factor" image copyright Biserka; "Musical Revolutions" image copyright Jeff deVries, iStockphoto; "Psych!" image copyright Janne Ahvo, iStockphoto.

Images in chapters "Mmmm, mmmm!," "One of a Kind — Platypus," "One of a Kind — You," "Snack Attack — Afternoon Munchies," and "Your Brain on Chocolate" are all copyright iStockphoto.

Images in chapters "High Moon," "Mad Scientists," "The First Astronauts," and "The Men on the Moon" are all courtesy NASA.

Images in chapters "Einstein's Brain: Lost & Found," "To Stick or Not to Stick," "Racing into the Future," "Write On!," "Computer Firsts," "Sun Gazing," "Paws 'n Claws," "IFOs: Identified Flying Objects," "Underwater Rainforests," "Digital Anatomy," "GermYnation: On the Job," "FAT Words," "What is . . . ?," "The Cat Connection," "The Wolf in Dog's Clothing," "Snack Attack: Movie Time," "The Car Kings," "Hot Chicken," "Survival 101: Public Restrooms," "The Sounds of Science," "GermYnation: Home Invasion," "Planet of the Cars," "Material World," and "Top Secret" are copyright Eva Everything.

Cover images: "Smoking gun," copyright Alan McCredie, iStockphoto; "International currency," copyright Joel Blit, iStockphoto; "Blue moon eclipse," copyright iStockphoto; "Shampoo dog," copyright Eric Isselée, iStockphoto; "Camembert," Tim Walton, iStockphoto.